T0348372

ADVANCES IN INTERNATIONAL ACCOUNTING

Volume 5 • **1992**

EDITORIAL BOARD

ADVANCES IN INTERNATIONAL ACCOUNTING

A Research Annual

Editor: **KENNETH S. MOST**
College of Business Administration
Florida International University

VOLUME 5 • 1992

 **JAI PRESS INC.**

Greenwich, Connecticut   *London, England*

CONTENTS

PART VI. INTERNATIONAL GOVERNMENTAL ACCOUNTING

LIST OF CONTRIBUTORS

Carlos Alcerreca-Joaquin

College of Business
University of Houston
United States

Daniel P.S. Asechemie

Department of Accounting
University of Port Harcourt
Nigeria

Aad D. Bac

Department of Accounting
Tilburg University
The Netherlands

Robert Bloom

School of Business
John Carroll University
United States

Susan C. Borkowski

Department of Accounting
LaSalle University
United States

Michael L. Costigan

Department of Accounting
Southern Illinois University
    at Edwardsville

Timothy S. Doupnik

College of Business Administration
University of South Carolina
United States

Jayne Fuglister

Department of Accounting
Cleveland State University
United States

Ali Mohamed Ibrahim

Department of Education
Sultan Qaboos University
Oman

Jeffrey Kantor

Department of Accounting
University of Windsor
Canada

Rifaat Ahmed Abdel Karim

Department of Accounting
Gezira University
Sudan

Eero Kasanen

Department of Accounting
Helsinki School of Economics
Finland

Juha Kinnunen

Department of Accounting
Helsinki School of Economics
Finland

James C. Lampe

College of Business Administration
Texas Tech University
United States

Jyrki Niskanen

Department of Accounting
Helsinki School of Economics
Finland

M.H.B. Perera

Department of Financial Accounting
Massey University
New Zealand

M.J. Pratt

School of Management Studies
University of Waikato
New Zealand

A.R. Rahman

Department of Financial Accounting
Massey University
New Zealand

Ahmed Riahi-Belkaoui

Department of Accounting
University of Illinois at Chicago
United States

Norlin G. Rueschhoff

Department of Accountancy
University of Notre Dame
United States

Stephen B. Salter

Department of Accounting
Texas A&M Universtiy
United States

Philip H. Siegel

College of Business
University of Houston
United States

Steve G. Sutton

Faculty of Management
University of Calgary
Canada

Pochara Theerathorn

College of Business and Economics
Memphis State University
United States

G.D. Tower

Commerce Programme
Murdock University
Australia

Pontus H. Troberg

Department of Accounting
Finnish Institute of Management
Finland

K.A. Van Peursem

Department of Accounting
and Finance
University of Waikato
New Zealand

INTRODUCTION

International accounting scholars feel gratified by any expression of interest in their specialty. We were pleased, therefore, to encounter a report of an interview given by Arthur Wyatt, chairman of the International Accounting Standards Committee, to Professors Kieso and Weygandt, which prefaces the latest edition of their book (1992).

In answer to the question "Does the United States have the best standards in the world?" Wyatt replied "[W]e have the most detailed disclosure standards and the most precise accounting standards. I wouldn't necessarily equate those two things to being the best." Obviously Wyatt is not familiar with the accounting standards of other countries, for example, France and Germany, which have extremely detailed and precise disclosure and accounting rules, although not in the same form as U.S. generally accepted accounting principles (GAAP). Every part of a German financial report responds to a specific provision of the law, and U.S. promulgated GAAP cannot compare in coverage with the French *Plan Comptable Général.*

Without having immersed oneself in the literature on international accounting, or having practiced accounting in a number of foreign countries, it is impossible to be aware of their regulations governing accounting and financial reporting. In some countries, the shorter the accounting textbook, the more precise the accounting standards. Indeed, the prolix, repetitive, and verbose nature of U.S. GAAP leads many (not Wyatt) to assume that "biggest is best."

This is an example of oversimplification, which characterizes much of accounting research at the present time. Oversimplification is to be distinguished from abstraction. Problems are usually observed in complex situations and cannot be analyzed, far less solved, without a degree of simplification. Abstraction simplifies by constructing a model that correctly represents the object being studied, yet omits features that are believed to be insignificant in the context of the investigation. Thus, early economic researchers in Western Europe abstracted from time, because there could be only one harvest a year in its farm-based economies. By the nineteenth century, however, this abstraction had become oversimplification, as the time period of production became a critical feature of industrial economies.

Oversimplification is encountered in the literature on international accounting that generally discusses harmonization and the influence of environmental factors, particularly culture, on accounting and financial reporting. It has been forced on many researchers by their ignorance of foreign languages, so that they are obliged to work with second- and third-hand sources in contravention of the first principle of research: Never observe indirectly that which can be observed directly. It is nurtured by ethnocentrism, manifested in the inability to conceive of reasons why others may do things differently. Questionnaires designed to identify differences fail to mention practices not found in the Anglo-Saxon world, such as:

- a firm's capital appearing in the balance sheet as an asset (remember General Electric's acquisition of Utah);
- separate disclosure of assets in use and assets not in use (the Dutch do this);
- disclosing how much depreciation expense is attributable to tax factors and how much is a business expense (Germany);
- amortizing purchased intangibles over the life specified for the capital budgeting decision (required under the French *Plan*);
- consolidated financial statements that omit those of the parent (Shell);
- proportional (pro rata) consolidation (this is commonly used for corporate joint ventures in Europe, but not approved by U.S. GAAP);
- contingent liabilities quantified and reported in balance sheet totals (guarantees under acceptances and endorsements of some European banks);
- inclusion of tangible and intangible assets under one balance sheet heading such as "fixed assets" (very common outside the United States); and
- separate audit reports on related party transactions or product costs (the former in France, the latter in India).

Perhaps the most pervasive form of oversimplification in harmonization research is reflected in the phrase "accounting system" which is the established term for accounting and financial reporting in a country (AlHashim and Arpan 1988, p. 15; Evans, Taylor, and Holzmann 1988, p. 19). This shorthand method of expression may embrace:

- accounting systems properly so-called, such as the *comptabilité par decalque* (copywrite bookkeeping) that is widely used in France and Switzerland;
- commercial laws that prescribe certain accounting records, such as the French *Code de Commerce*;
- company laws that govern accounting and auditing, such as the U.K. Companies Acts;
- legislation regulating banks, insurance companies, public utilities, state enterprises, and other specialized entities;
- ministerial regulation, such as the Business Deliberation Council in Japan;
- securities regulation, such as the Securities and Exchange Commission in the United States;
- income tax laws, which regulate accounting in tax conformity countries;
- pronouncements of professional associations, such as the Canadian Institution of Chartered Accountants, incorporated by statute into the company laws of the country;
- GAAP promulgated by bodies specially created for the purpose, like the Financial Accounting Standards Board in the United States;
- complex interaction between the public and private sectors, which characterizes accounting standard-setting in Australia;
- decision of law courts, such as the cases that govern auditing in the United Kingdom or the accounting court in The Netherlands; and
- the financial reports of large, particularly multinational, corporations which tend to serve as innovators and pioneers in their countries of domicile.

It is clear that each of these is itself an object of investgation, and comparing the "accounting systems" of different countries contravenes the second principle of research: Never study jointly that which can be separated, or in Descartian terms, divide every problem into as many parts as are necessary in order to solve it.

I cannot claim that the research published in the pages of *Advances in International Accounting* can withstand the accusation of oversimplification, but in my correspondence with authors, and through the editing process, I try to guard against it. For this I am also grateful to the editorial board, whose cooperation makes this publication possible.

Miami, Florida **Kenneth S. Most**, Editor

REFERENCES

AlHashim, D.D., and J.S. Arpan. 1988. *International Dimensions of Accounting*. 2nd ed. Boston: PWS-Kent.

Evans, T.G., M.E. Taylor, and O. Holzmann. 1988. *International Accounting and Reporting*. Boston: PWS-Kent.

Kieso, D.E., and J.J. Weygandt. 1992. *Intermediate Accounting*. 7th ed. New York: Wiley.

PART I

INTERNATIONAL ACCOUNTING THEORY

THE RELATIONSHIP BETWEEN
LEGAL SYSTEMS AND
ACCOUNTING PRACTICES:
A CLASSIFICATION EXERCISE

Stephen B. Salter and Timothy S. Doupnik

ABSTRACT

This paper empirically examines the relationship between worldwide legal systems and accounting practices. A hierarchical classification of legal systems was developed based on the comparative research of David and Brierley and clusters of countries with similar accounting systems were identified using current data on financial reporting practices and hierarchical cluster analysis. A comparison of the two classification schemes supports the hypothesis that a country's legal system is a significant predictor of membership in a particular accounting cluster.

Advances in International Accounting,
Volume 5, pages 3-22.
Copyright © 1992 by JAI Press Inc.
All rights of reproduction in any form reserved.
ISBN: 1-55938-415-8

3

INTRODUCTION

A growing body of literature exists on the classification of accounting systems worldwide (see Meek and Saudagaran 1990, for a review of this literature). Classification studies attempt to group countries according to similarities in financial reporting practices. A practical benefit of classification may be that it provides information to those involved in accounting harmonization on the systematic differences that exist among groups of countries, which need to be addressed. A time series of classifications based on actual practices also can provide information on the extent to which harmonization is being realized. Classification might also prove useful to multinational corporations with foreign audiences of interest outside their own group in identifying those situations where additional disclosures might be warranted (Meek and Saudagaran 1990). The corollary to this is that classification studies provide information to international investors as to when care should be exercised in using financial statements of companies from countries outside their own cluster (Nair 1982).

A number of environmental factors, such as level of economic development, legal system, and source of corporate funds, have been hypothesized as influencing a country's accounting system. Identification of those environmental factors most significantly related to countries' accounting systems could provide guidance as to how harmonization should proceed. For example, knowledge that level of economic development significantly affects accounting practices might suggest that harmonization efforts should concentrate on countries with similar levels of economic development.

Several studies have attempted to empirically examine the relationship between nonaccounting environmental variables and accounting systems (Frank 1979; Nair and Frank 1980; Goodrich 1986). Each of these studies suffers from methodological problems primarily related to the use of subjective data and the use of inappropriate statistical analysis techniques. In addition, each of these studies was based on data from the year 1979 or earlier. Recent changes in accounting practices in various countries (perhaps most notably in the European Community) may mean that the results of previous studies are now out of date.

The objectives of the current study are to (1) develop a current classification of international accounting systems using multiple-source data and appropriate clustering techniques, and (2) examine the relationship between the emergent clusters of countries and a single environmental variable: type of legal system.

RELEVANT LITERATURE

Numerous papers have been written on the relationship between accounting and its environment and a host of factors influencing a country's accounting system have been hypothesized (see, e.g., Seidler 1967; Mueller 1968; Radebaugh 1975; AAA 1977; Choi and Mueller 1984; Arpan and Radebaugh 1985). Meek and Saudagaran (1990, 150-51) indicate that there is general agreement that the following factors influence accounting development: (1) legal system, (2) nature of the relationship between business enterprises and providers of capital, (3) tax laws, (4) inflation levels, and (5) political and economic ties. The current study focuses on legal system as a determinant of accounting system.

Relationship Between Legal and Accounting System

A country's laws can impact accounting practices in a number of ways. Companies' acts, found in many countries, contain laws regulating business activities including the keeping of accounting records and publication of financial statements. Tax laws in many countries specify accounting procedures to be used in the area of financial reporting as well as taxation. The issue is not whether accounting rules are in some fashion legislated, but rather the extent to which they are determined by law. Formal codification of accounting standards is typically found in Roman or Code law countries, whereas accounting rules are established in a nonlegalistic manner in Common law countries (Choi and Mueller 1984, 41).

The degree to which accounting rules are legislated can impact the nature of the accounting system. Meek and Saudagaran (1990, 150) argue that in Code law countries, laws stipulate minimum requirements and accounting rules tend to be highly prescriptive and procedural. Compliance with the letter of the law is expected. In Common law countries, on the other hand, laws establish limits beyond which it is illegal to venture and within those limits experimentation is encouraged and judgment is required. Within this institutional framework accounting rules tend to be more adaptive and innovative.

In a critique of accounting classification studies, Most and Salter (1990) suggest that past classification schema have lacked empirical referents, and attempt to analyze "accounting systems" as if the phrase was a unitary construct rather than an amalgam of different kinds of laws, customs, and professional rules. Because accounting is legislated at least to some degree in every country of the world, an initial step in understanding differences in accounting systems should be a comparison of the legal system under which accounting rules are developed. They suggest an avenue of research in which accounting "laws" are mapped onto accounting practices to determine the extent to which existing accounting practices are a function of law. They imply that most accounting practices can be explained in this manner. This might be a fruitful avenue of research. The logical extension of this would be to then concentrate on the relationship between those practices unexplained by legislation and other environmental factors.

The current study represents a first step in this direction. Rather than tracing individual accounting practices to their basis in specific laws and regulations within a country, which would be an enormous task for any single country, this study examines the relationship between an international classification of legal systems based on work conducted by legal scholars and a classification of international accounting systems based on a statistical analysis of current practices. The results might provide preliminary evidence on the importance of legal system as a determininant of worldwide accounting systems and provide support for Most and Salter's (1990) hypothesis that substantial differences in accounting systems can be explained by reference to a country's laws. As such, the results could have implications for future research on the environmental determinants of accounting systems.

Empirical Accounting Classification Studies

A number of studies have attempted to classify countries into groups based on the data bases of financial reporting practices compiled by Price Waterhouse (PW) International (1973, 1975, 1979; Da Costa, Bourgeouis and Lowson 1978; Frank 1979; Nair and Frank 1980; Nair 1982; Goodrich 1986). In addition, each of these studies used factor analysis as a clustering tool.

These studies have been criticized both for the data used and the statistical methods employed to identify commonalities between countries. Nobes (1981) criticizes the use of PW data for classification because of outright errors, swamping of important information by trivial data, exaggeration of certain intercountry differences, and failure to differentiate desired practices from reality. Nobes (1987) also criticized the empirical studies for a lack of a model against which the reliability of the statistical results could be evaluated. In addition to these criticisms, the empirical studies are flawed in that the use of factor analysis as a grouping tool is "an extreme perversion of the method" (Stewart 1981, 51).

In addition to developing groupings of countries by financial reporting practices, Frank (1979), Nair and Frank (1980), and Goodrich (1986) tested the ability of factors generated from environmental variables to predict those groupings. Frank (1979) and Nair and Frank (1980) examined three types of factors: culture, economic structure, and trading blocks. Goodrich (1986) considered political systems and attitudes, socioeconomic factors, and membership in international organizations. None of these studies examined the relationship between legal system and accounting practices.

While Frank (1979) and Nair and Frank (1980) found a number of factors to be statistically significant in predicting observed groupings, the results may be questioned given that the ratio of data points to predictor variables was in all cases below the norm of five to one (e.g., Nair and Frank [1980] used 18 variables to discriminate among 46 countries). Moreover, a number of statistically significant predictors lack intuitive appeal. For example, it is difficult to understand why the trading block of Mexico, Panama, and Trinidad and Tobago should be one of Nair and Frank's (1980) six key discriminant variables of worldwide accounting disclosure systems.

In addition, the reliability of the dependent variables used in each of these studies is questionable in that they were developed using PW data and factor analysis. The accounting principles groups developed by Goodrich (1986) are especially troublesome. For example, given the results of prior research (e.g., Seidler 1967; Frank 1979; Nobes 1984), an accounting cluster that includes Brazil, Norway, Australia, and Japan appears improbable. Frank (1979), Nair and Frank (1980), and Goodrich (1986) make no attempt to validate the reliability of the country groupings obtained from their statistical analysis.

RESEARCH QUESTION AND METHODOLOGY

The major question addressed in this study is: To what extent can clusters of countries with similar accounting practices be predicted by the countries' legal systems? The null hypothesis tested with regard to this question is:

H_0: A classification of worldwide legal systems is not a significant explanator of clusters of accounting systems worldwide.

To test this hypothesis it was necessary to develop (1) a classification of countries by legal system and (2) a classification of countries by accounting practice. The former was developed through examination of the comparative study of legal systems conducted by David and Brierley (1985).

The latter was developed through statistical analysis of data on countries' accounting practices specifically gathered for this study. The current study attempts to improve on previous accounting classification efforts by (1) obtaining current information on accounting practices from multiple sources, rather than a single source such as the PW surveys, and (2) using hierarchical cluster analysis, rather than factor analysis, as the clustering tool. In addition, the reliability of the results of the cluster analysis are partially evaluated by comparing them with hierarchical classifications of financial reporting systems developed by Nobes (1984) and Berry (1987).

The relationship between the classification of accounting systems and the classification of legal systems is then examined by calculating "hit ratios" using type of legal system as the predictor (independent) variable and accounting cluster as the criterion (dependent) variable. The significance of the hit ratios is evaluated using heuristics suggested by Hair, Anderson, and Tatham (1987).

A CLASSIFICATION OF LEGAL SYSTEMS

David and Brierley (1985) have identified three major families of legal systems: Romano-Germanic (Code) law, Common law, and Socialist law. (As the Soviet Union and its former satellites are not included within the scope of the current study, discussion of Socialist law is not of immediate interest.)

Romano-Germanic law, as the name implies, is European in origin and is characterized by rules of law being formulated by legal scholars based on ideas of justice and morality. These rules of law form a general framework within which any legal issue can be resolved. Most of the countries in the Romano-Germanic family codified their laws during the nineteenth and twentieth centuries adopting the organizational framework of the French Napoleonic Codes written during the period 1804-1811 (hence, the common name for this family as Code law countries). All of the countries of continental Western Europe belong to the Romano-Germanic family. This legal system has spread to other parts of the world through colonization and through voluntary reception in those non-European countries where the need to modernize or the desire to Westernize has led to the penetration of European ideas. David and Brierley suggest that while still a part of the Romano-Germanic family, the system is different in each of the non-European branches of the family as the system has confronted traditional laws (Middle East, Asia, Africa) or adapted to local conditions (Latin America). Indeed, within the European branch of the family, unique subsets exist (Germany and Sweden/Finland).

The other major legal system in the world today, Common law, is English in origin. In contrast to the Romano-Germanic approach, Common law was developed primarily by judges as they resolved specific disputes. Common law legal rule seeks to provide a solution to the case at hand rather than formulate a general rule of conduct for the future. Common law also has spread to other parts of the world primarily through expansion of the British Empire. David and Brierley suggest that alterations in the basic system vary in importance and nature according to the strength of ties with England, geographical conditions, the influence of indigenous civilizations, and other factors (1985, 397). Differences in the nature of the Common law system in Europe (the United Kingdom and Ireland) and outside Europe are most likely to exist in Islamic countries, India, and North America. In fact, the United States and Canada are thought to "enjoy a largely autonomous place within the family" (David and Brierley 1985, 25).

Although David and Brierley do not themselves develop a hierarchical classification of the world's legal systems, based on their comparative analysis, such a hierarchy appears to exist. Based on their work, a hierarchical classification of legal systems was developed for the 50 countries for which data on financial reporting practices were gathered (discussed below). The resulting hierarchy is presented in Figure 1.

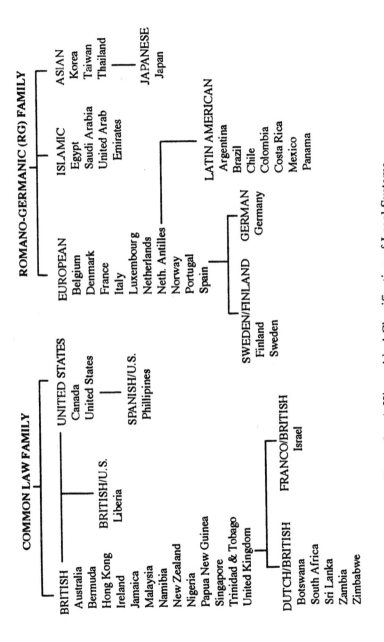

Figure 1. A Hierarchical Classification of Legal Systems

9

Common Law Family

The Common law legal system is used in the United Kingdom and all of the countries that have at some time been a British possession or protectorate. There are two main branches of this family of legal system: British and United States.

Due to its early break from the British empire, the legal system in the United States has evolved to be distinctly different from that in Britain. Although not compared in detail, David and Brierley (1985, 25) suggest that the legal system in Canada is also more like that in the United States than in the United Kingdom.

Within the British branch of the family, two hybrid groups exist. The first consists of countries which, prior to their annexation by Britain, "belonged to the Romano-Germanic family by reason of their Dutch colonisation" (David and Brierley 1985, 77). The legal system in these countries was influenced by British rule such that they are now considered to have a mixed legal system: Romano-Germanic/Common law. The second group consists of Israel where, "because of the British mandate, the influence of the Common law has largely supplanted that of the Franco-Ottoman law formerly in force" (1985, 78). Because these countries have most recently been influenced by British rule, we classify them as Common law countries.

The Philippines represent a mixed legal system within the United States branch of the Commmon law family. Spanish colonization brought the Philippines into the Romano-Germanic legal family. "Fifty years of occupation by the United States have nevertheless introduced new elements, making Philippine law a mixed system" (1985, 79).

Liberia is a Common law country, but it cannot be clearly classified as a member of the British or United States branch. David and Brierley (1985, 560) indicate that when Liberia was founded, Common law and practices of both English and American courts were declared applicable. Liberia has thus been placed in a separate British/United States subbranch of the Common law family.

Romano-Germanic Family

Based on David and Brierley's work, five major branches of the Romano-Germanic (RG) family appear to exist: European, Latin American, Asian, Islamic (Arab), and Black African. (There are no RG-African countries included in this study.) In addition to the main body of the European branch, two subbranches exist: Sweden/Finland and Germany.

The Nordic countries (Denmark, Finland, Iceland, Norway, and Sweden) each adopted a single code of law prior to the writing of the Napoleonic Codes in France in the early 1800s. Those earlier single codes have since disappeared in Denmark, Iceland, and Norway, supplanted by codes organized along the lines of the Napoleonic codes (David and Brierley 1985, 112). Unique among the RG countries, Sweden and Finland continue to operate under a single code that dates to 1734. We classify Sweden/Finland as a subbranch of the main RG-European branch of the RG family.

In Germany, the structure of the law is different from other members of the RG family. David and Brierley question whether the German legal system is different enough to represent a break from the RG family. After further analysis they conclude that although different, the difference is not great enough to "compromise the unity of the Romano-Germanic family" (1985, 91-2). Because of its unique characteristics, we have placed Germany in a separate subbranch of the RG-European branch of the RG family.

The RG legal system is used in all of Latin America by virtue of Spanish, French, Dutch, and Portuguese colonization. David and Brierley raise the question of whether the system in Latin America represents a secondary grouping within the RG family (1985, 34). They question "the extent to which the laws of America, which grew to maturity in conditions very different from those prevailing in Europe, have developed original characteristics when compared to the European laws of the Romano-Germanic family" (p. 75). The question is never specifically answered. However, the fact that the question is raised suggests that although very similar to the RG-European legal system, a separate RG-Latin America legal system might exist. We have placed such a subbranch under the main (European) branch of the RG family.

Several Arab states (including Egypt) have codified their laws patterned on the French model. However, in most of these countries the code instructs judges to fill any gaps in the law according to the principles of Islamic law, thus creating a mixed RG-Islamic legal system. The situation in the countries of the Arabian peninsula is less clear. David and Brierley indicate that this area has undergone little RG influence up to the present time. They state that "it cannot not now be said whether the English and American economic influence ... will prevail or whether closer links with the Egyptian and Arab worlds will orient the laws of these countries towards at least a partial adherence to the Romano-Germanic family" (1985, 78). Given the underlying Islamic legal tradition in Saudi Arabia and the United Arab Emirates, which is consistent with other countries of the Arab world, we have tenuously placed these two countries as members of the RG-Islamic subbranch of the RG family rather than in one of the branches of the Common law family.

David and Brierley indicate that most of the non-British Asian countries have adopted codes, thereby rejecting their traditions "in favour of Romano-Germanic law as the basis for social relations" (1985, 516-7). This group includes Japan, South Korea, Taiwan, and Thailand. However, Japan is different from the other countries of this group because since 1945 "an Anglo-American influence has been at work on, and is sometimes in competition with, the Romanist influence" (1985, 540). Thus, Japan has been placed in a separate subbranch of the RG-Asian branch of the RG family.

DEVELOPMENT OF A CLASSIFICATION OF ACCOUNTING SYSTEMS

To obtain data on current international accounting practices for the purpose of developing a classification of accounting systems, a survey instrument based on the 1979 PW survey was developed. To avoid the criticism that the PW surveys swamp important issues with trivial ones, the current instrument included PW propositions which were identified as important by virtue of their being addressed in International Accounting Standards 1—29. In addition, accounting issues not included in the PW survey were added to obtain data on various issues addressed by the IASC and FASB since 1979. Propositions were selected to ensure that no important financial reporting issue was omitted or overweighted. The questionnaire initially consisted of 114 financial reporting practices.

The preliminary questionnaire was reviewed by an expert panel of eight academicians, ten practicing accountants, and two international investment analysts. The persons were familiar through academic training or experience with accounting principles in countries of Europe, North America, South America, the Caribbean, Australasia, and Africa. Panel members were asked to: (1) rate each practice on a three-point scale indicating the importance of each practice, (2) suggest alternate practices of key importance, and (3) indicate any items that were confusing or might be difficult to interpret. Based on panel input a number of alterations were made and, to improve the probability of an adequate response, the instrument was shortened to those 100 practices with the highest ratings.

A pretest was carried out in three countries (12 participants) to test consistency of responses across respondents (within countries). No significant differences were found at the .05 level.

The revised instrument was sent to the managing partners of offices of international public accounting firms in each of the countries in which at least two firms had offices. The firms surveyed were Arthur Andersen, Coopers and Lybrand, Deloitte and Touche, Ernst and Young, KPMG, Laventhol and Horwath, Panell Kerr Foster, and Price Waterhouse. In most cases the instruments were mailed directly from the participating firm or were accompanied by a letter of support from a senior partner in the firm's New York or Toronto office. Participants were asked to indicate on a 0-100 scale the percentage of economically significant entities following specific financial reporting practices. One of the problems faced by earlier researchers using PW data was the need to subjectively transform the categorical data gathered into some kind of a continuous scale for input into factor analysis. That problem was avoided in this study. In addition, the participants were specifically requested to provide information on actual reporting practices and not on officially pronounced accounting rules as was the case in the PW surveys. Respondents were also specifically asked to limit their responses to "companies with which you have had prior audit or business contact." To assist respondents in assigning percentages, a verbal rating scale with suggested equivalent numeric scores was provided.

To avoid the problems of error and subjectivity leveled against the PW data, only those countries for which at least two, consistent responses were obtained were included in the analysis. Single responses were received from 30 countries and they were dropped from the data set. To ensure that a consensus of opinion was reached by respondents within a country, tests were conducted to identify and delete respondents whose pattern of responses was significantly different from others in the same country. Analysis of variance identified 36 individuals whose responses were significantly different from other respondents from the same country and they were dropped from the data set. This resulted in eight additional countries being excluded from further analysis. After initial clustering, another eight countries identified as outliers were eliminated from further analysis as suggested by Punj and Stewart (1983). The final data set consisted of 174 responses distributed across 50 countries. These 50 countries include 19 of the 20 countries classified in Nobes (1984) and 28 of the 37 noncommunist countries classified by Berry (1987).

In summary, the data set contained a mean of 3.46 responses per country from respondents who had a mean audit experience of 16.7 years. Respondents indicated that their responses were based on a mean of at least 40 companies with which they had prior audit or other business contact. The reduced data set was used as input into cluster analysis to identify clusters of countries with similar accounting practices.

Cluster Analysis

In using cluster analysis, there are initially two decisions to be made: the method of clustering (hierarchical vs. nonhierarchical), and the algorithm to be used. Because prior deductive work (Nobes 1983, 1984; Berry 1987) suggests a multilevel structure to international accounting systems, hierarchical cluster analysis was used, letting the data reveal the number of significant clusters through various heuristics discussed below.

Two algorithms—average linkage method and Ward's method—have been cited as superior to others (Punj and Stewart 1983). As no consensus exists as to which of these two is best, both were used initially and the superior of the two was determined using the validation technique developed by McIntyre and Blashfield (1980). The degree of agreement between the nearest centroid assignment of the holdout sample and the results of a cluster analysis of the holdout sample was found to be 100% for the average linkage method, an indication of the stability of this solution. The degree of agreement was only 60% for Ward's method, an indication of the instability of this solution. On the basis of these results all further analysis was based on the results obtained from the average linkage method of clustering.

The number of significant clusters was determined by examining pseudo F and pseudo t^2 statistics. The pseudo F statistics were examined for local and maximum peaks (SAS 1985). Pseudo t^2 statistics were examined for breaks or rapid drops in value (Hair et al. 1987). As can be seen from Figure 2, the best (global) solution using Pseudo F peak and t^2 drop occurred at two clusters with a strong local solution at nine clusters. The results of each of these cluster solutions are described below and briefly compared with Nobes (1984) and Berry (1987) as a partial check on the reliability of the emerging solutions.

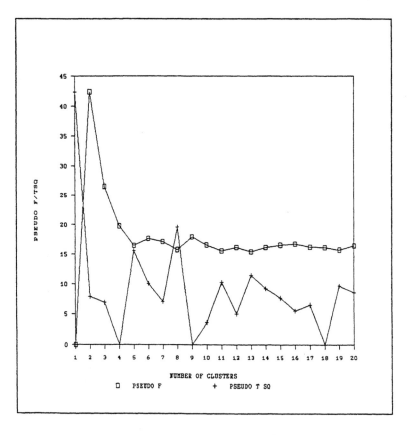

Figure 2.   Pseudo F and Pseudo t^2 Statistics

Two Cluster Solution

The two-cluster solution (presented in Table 1) resulted in two clusters of countries which generally correspond to Nobes' "micro-based" and "macro-uniform" classes of financial reporting systems. The micro group includes all present and former British Commonwealth countries included in the sample, the United States, and the Netherlands and Netherlands Antilles. It also includes the Philippines, Taiwan, Israel, and Luxembourg. Japan, Mexico, and Chile are not found, as hypothesized by Nobes (1984), to be members of the micro-based class of countries. Japan and Chile as members of the macro-uniform class of countries is consistent with Berry (1987). The macro group includes all continental European countries (excluding Luxembourg and the Netherlands) and countries in Latin America, the Arab World, and non-British Asia and Africa.

Nine Cluster Solution

The nine-cluster solution (presented in Table 2) splits the micro group into two subgroups:

1. Dutch/British Commonwealth group (C1), and
2. a United States/Canada group (C2).

The macro group split into seven subgroups:

1. Southern European dominated group (C3),
2. Arab dominated group (C4),
3. Latin American group (C5),
4. Sweden/Finland (C6),
5. Costa Rica (C7),
6. Japan (C8), and
7. Germany (C9).

Comparing the six families hypothesized in Nobes (1984) with the nine clusters emerging in the current study, seven countries (the Netherlands, Japan, the Philippines, Mexico, Chile, Brazil, and Belgium) are classified differently. Comparing the seven families hypothesized in Berry (1987), six countries are classified differently (the Netherlands, Japan, the Philippines, Mexico, Colombia, and Belgium).

The emergence of separate British Commonwealth and United States/Canada groups is consistent with Nobes' hierarchy, as is the emergence of a predominantly Southern European group, with Sweden and Germany forming separate groups. Nobes (1984) did not consider the Arab world in his

Table 1.   Two-Cluster Solution

Cluster 1 (*Micro*)	Cluster 2 (*Macro*)
(N) Australia	(B) Argentina
(B) Bermuda	(N) Belgium
Botswana	(N) Brazil
(N) Canada	(N) Chile
Hong Kong	(B) Colombia
(N) Ireland	Costa Rica
Israel	Denmark
(B) Jamaica	Egypt
Luxembourg*	Finland
(B) Malaysia	(N) France
Namibia	(N) Germany
(N) Netherlands*	(N) Italy
Netherlands Antilles*	(N) Japan
(B) Nigeria	Korea (S)
(N) New Zealand	Liberia*
(N) Philippines	(N) Mexico
Papua New Guinea	Norway
(N) South Africa	Panama
(B) Singapore	Portugal
Sri Lanka	Saudi Arabia
Taiwan*	(N) Spain
(B) Trinidad and Tobago	(N) Sweden
(N) United Kingdom	Thailand
(N) United States	United Arab Emirates
Zambia	
(B) Zimbabwe	

Notes: (N) Denotes countries included in Nobes (1984).
　　　　　(B) Denotes countries included in Nobes (1984) and Berry (1987).
　　　　　　* Denotes countries misclassified by legal system.

classification, nor did he hypothesize a separate Latin American family. However, the emergence of a Latin American group is consistent with Berry (1987).

In contrast to both Nobes (1984) and Berry (1987), The Netherlands does not appear different enough from the British Commonwealth countries to form a separate group. This is consistent with empirical tests conducted by Nobes (1984). Japan was included in the United States influence family in Nobes (1984) and a group with Germany in Berry (1987). Nobes (1984) admits that classification of Japan is difficult because of the nature of its accounting development, that is, medieval bookkeeping overlaid first by German and United States practices. The current results suggest that this hybrid results in a unique accounting system which is not highly related to either its German or United States counterparts. Neither the Philippines nor Mexico emerge in the United States group as hypothesized by both Nobes (1984) and Berry

Table 2. Nine-Cluster Solution

C1	C2	C3
(N) Australia	(B) Bermuda*	(B) Colombia*
Botswana*	(N) Canada	Denmark
Hong Kong	Israel*	(N) France
(N) Ireland	(N) United States	Korea*
(B) Jamaica		(N) Italy
Luxembourg*		Norway
(B) Malaysia		Portugal
Namibia		(N) Spain
(N) Netherlands*		
Netherlands Antilles*		
(B) Nigeria		
(N) New Zealand		
(N) Philippines*		
Papua New Guinea		
(N) South Africa*		
(B) Singapore		
Sri Lanka*		
Taiwan*		
(B) Trinidad and Tobago		
(N) United Kingdom		
Zambia*		
(B) Zimbabwe*		

C4	C5	C6
(N) Belgium	(B) Argentina	Finland
Egypt	(N) Brazil	(N) Sweden
Liberia*	(N) Chile	
Panama	(N) Mexico	
Saudi Arabia		
Thailand*		
United Arab Emirates		

C7	C8	C9
Costa Rica*	(N) Japan	(N) Germany

Notes: (N) Denotes countries included in Nobes (1984).
(B) Denotes countries included in Nobes (1984) and Berry (1987).
* Denotes countries misclassified by legal system.

(1987). Apparently United States influence on the accounting practices in those two countries is not as great as hypothesized. The emergence of Colombia in the Southern European group and Belgium in the Arab-dominated group was unexpected.

The results of the clustering procedures used in this study are largely consistent with hierarchical classification schemes developed by Nobes (1984) and Berry (1987). The differences that exist between the empirical results and the hypothesized classifications may be a result of several factors including the natural evolution of accounting systems since the classifications were developed and overweighting of specific factors in subjectively classifying individual countries. We felt that the groupings obtained were consistent enough with the hypothetical classifications (and differences are likely to be due to faulty deduction) to use them in subsequent testing of the legal system hypothesis.

TEST OF THE LEGAL SYSTEM HYPOTHESIS

To test the hypothesis that a country's legal system predicts the accounting cluster to which that country belongs, hit ratios were calculated at the two-cluster and nine-cluster levels. While no direct heuristic exists to evaluate the hit ratio, a good comparison might be the maximum chance and proportional chance models used in assessing the predictive accuracy of discriminant functions (Hair et al. 1987, 89-90). Maximum chance is computed by dividing the number of countries in the largest accounting cluster by the total number of countries. Proportional chance is computed by summing the squared proportions in each accounting cluster. Hair et al. (1987, 90) suggest that for a discriminant function to be significant, its hit ratio must exceed 1.25 times the greater of the maximum chance or proportional chance ratios. In this case the maximum chance ratio was greater for both levels. The maximum chance heuristics reported equal 1.25 times the maximum chance ratio.

At the two-cluster level, legal system family (Common or RG) correctly classifies class of accounting system (micro or macro) for 45 of 50 countries for a hit ratio of 90% (see Table 1). This greatly exceeds the maximum chance heuristic of 65%, indicating that at the class level the type of legal system is a significant predictor of accounting system. The misclassified countries are Luxembourg, the Netherlands and Netherlands Antilles, and Taiwan, which have micro accounting systems but belong to the RG family of legal system, and Liberia, which is a Common law country but has a macro accounting system.

A comparison of the nine accounting clusters in Table 2 with the 13 groups of countries in Figure 1 shows that 31 countries are correctly predicted for a hit ratio of 62% (see Table 2). This compares favorably with a maximum chance heuristic of 55%, indicating that legal system is a significant predictor of accounting cluster at the nine-cluster, family level. Thus, the null hypothesis may be rejected at both the two-cluster and nine-cluster levels.

There are two major systematic differences between the classification of legal systems and the accounting clusters at the nine-cluster level. The first is that within the macro class of accounting systems a separate Asian accounting cluster does not emerge. Instead, the countries of the Asian branch of the RG family emerge in three different accounting clusters (C1, C3, and C4).

The second is that within the micro class of accounting systems only two families emerge (C1 and C2), whereas within the Common law family of legal systems six groups emerge. If the hybrid subbranches of the Common law family were collapsed into the two major branches (British and United States) as shown in Table 3, 36 countries (72%) would be correctly classified by accounting cluster as each of the members of the Dutch/British subbranch of the Common law family are found in the Dutch/British Commonwealth accounting cluster (C1). For those countries, further research might be worthwhile to determine whether the Common law legal system has completely replaced the Dutch system that might have at one time been in place.

Table 3. Classification of Legal Systems at Nine-Cluster Level

Common-British	Common-United States	RG-European
Australia	Canada	Belgium*
Bermuda*	Philippines*	Denmark
Botswana	United States	France
Hong Kong		Germany
Ireland		Italy
Israel*		Luxembourg*
Jamaica		Netherlands*
Liberia*		Netherlands Antilles*
Malaysia		Norway
Namibia		Portugal
New Zealand		Spain
Nigeria		
Papua New Guinea		
Singapore		
South Africa		
Sri Lanka		
Trinidad and Tobago		
United Kingdom		
Zambia		
Zimbabwe		

RG-Islamic	RG-Latin America	RG-Sweden/Finland
Egypt	Argentina	Finland
Saudi Arabia	Brazil	Sweden
United Arab Emirates	Chile	
	Colombia*	
	Costa Rica*	
	Mexico	
	Panama*	

RG-Asia	RG-Japan	RG-Germany
Korea*	Japan	Germany
Taiwan*		
Thailand*		

Note: * Denotes countries emerging in different accounting cluster.

DISCUSSION OF RESULTS

Results of this study provide broad support for the hypothesis that legal system is a significant predictor of accounting systems worldwide. The results clearly indicate a dichotomization of accounting systems consistent with the Common law/Romano-Germanic dichotomization of legal systems.

This could imply serious problems for worldwide accounting harmonization. As long as two fundamentally different legal systems exist in the world, harmonization across type of legal system may not be feasible. Note the results associated with members of the European Community. The Common law members emerge in the micro class of accounting system whereas most of the RG members emerge in the macro class. This is true even after implementation of the EC's 4th and 7th Directives in most member nations. The results indicate some hope for harmonization, however, in that two RG members of the EC—Luxembourg and the Netherlands—emerge in the micro accounting group. Additional research on accounting regulation in those two countries in particular might provide insight as to how the apparent linkage between legal system and accounting can be broken.

Although legal system accurately predicted accounting cluster at the nine-cluster level 62% of the time, that still leaves 38% of the countries whose accounting cluster could not be predicted by that factor alone. Future classification efforts should perhaps concentrate on countries such as the Netherlands, Israel, and Colombia to explain why these countries emerge in accounting clusters inconsistent with their membership in a particular branch of a legal family.

As with any empirical study, the results must be evaluated in light of several methodological limitations. Although great care was taken to avoid the problems that have been associated with using PW data for classification purposes, the data used in this study were also based on subjective assessments of accounting professionals and errors surely exist. It is possible that systematic biases across respondents from a specific country exist which our consistency filter was not able to control for. This is more likely to be true for those countries which were classified differently from Nobes (1984) or Berry (1987). We are especially uncomfortable with Belgium emerging in an Arab-dominated accounting family. An alternative method of collecting data on accounting practices might involve an examination of published financial statements of individual companies. Although we doubt that the same level of detail on accounting measurement practices can be gleaned from published annual reports, replication of this study using data from alternative sources might make a valuable triangulation contribution.

A second limitation relates to the development of the classification of legal systems. This classification was based on a single source of information and thus is in error to the extent that David and Brierley's (1985) analysis is faulty.

In addition, in several instances we were forced to make inferences as to how the legal systems of individual or groups of countries relate to others. However, we do feel that a relatively strong relationship does exist between legal and accounting systems and that tentative support has been provided for Most and Salter's (1990) assumption that substantial differences in accounting systems can be explained by reference to a country's laws.

REFERENCES

American Accounting Association. 1977. Report of the Committee on International Accounting Operations and Education. *Accounting Review* Supplement: 65—132.

Arpan, J., and L. Radebaugh. 1985. *International Accounting and Multinational Enterprises*, 2nd ed. New York: Wiley.

Berry, I. 1987. The need to classify worldwide accountancy practices. *Accountancy* (October): 90-91.

Choi, F., and G. Mueller. 1984. *International Accounting*. Englewood Cliffs, NJ: Prentice-Hall.

Da Costa, R., J. Bourgeois, and W. Lawson. 1978. A classification of international financial accounting practices. *International Journal of Accounting* (Spring): 73-85.

David, R., and J. Brierley. 1985. *Major Legal Systems in the World Today*. London: Stevens.

Frank, W. 1979. An empirical analysis of international accounting principles. *Journal of Accounting Research* (Autumn): 593-605.

Goodrich, P. 1986. Cross-national financial accounting linkages: An empirical political analysis. *The British Accounting Review* (Autumn): 42-60.

Hair, J., R. Anderson, and R. Tatham. 1987. *Multivariate Data Analysis with Readings*. New York: MacMillan.

McIntyre, R. and R. Blashfield. 1980. A nearest centroid technique for evaluating minimum variance clustering procedures. *Multivariate Behavior Research* 225-238.

Meek, G., and S. Saudagaran. 1990. A survey of research on financial reporting in a transnational context. *Journal of Accounting Literature* 145-182.

Most, K.S. and S.B. Salter. 1990. Classification research in international accounting and its relevance to european accounting harmonization. *Proceedings of the CGA/Concordia University Forum on Global Accounting*, Montreal.

Mueller, G. 1968. Accounting principles generally accepted in the U.S. versus those generally accepted elsewhere. *The International Journal of Accounting* (Spring): 91-104.

Nair, R. 1982. Empirical guidelines for comparing international accounting data. *Journal of International Business Studies* (Winter): 85-98.

Nair, R., and W. Frank. 1980. The impact of disclosure and measurement practices on international accounting classifications. *The Accounting Review* (July): 426-450.

Nobes, C. 1981. An empirical analysis of international accounting principles: A comment. *Journal of Accounting Research* (Spring): 268-270.

———. 1983. A judgemental international classification of financial reporting practices. *Journal of Business, Finance and Accounting* (Spring): 1-19.

———. 1984. An hypothesis and some tests of it. *International Classification of Financial Reporting*. London: Croom Helm.

———. 1987. Classification of financial reporting practices. Pp. 1-22 in *Advances in International Accounting*, Vol. 1, edited by K.S. Most. Greenwich, CT: JAI Press.

Price Waterhouse International. 1973. *Accounting Principles and Reporting Practices: A Survey in 38 Countries*. London: PWI.

_____. 1975. *Accounting Principles and–Reporting Practices: A Survey in 46 Countries.* London: PWI.

_____. 1979. *International Survey of Accounting Principles and Reporting Practices.* London: PWI.

Punj, G., and D. Stewart. 1983. Cluster analysis in marketing research: Review and suggestions for application. *Journal of–Marketing Research* (May): 134-148.

Radebaugh, L. 1975. Environmental factors influencing the development of accounting objectives, standards, and practices in Peru. *The International Journal of Accounting* (Fall): 39-56.

SAS Institute. 1985. *SAS User's Guide: Statistics, Version 5 Edition.* Cary, NC: SAS.

Seidler, L. 1967. International accounting—the ultimate theory course. *The Accounting Review* (October): 775-781.

Stewart, D. 1981. The application and misapplication of factor analysis in marketing research. *Journal of Marketing Research* (February): 51-62.

PART II

COUNTRY STUDIES

RECENT DEVELOPMENTS IN FINANCIAL REPORTING IN FINLAND

Pontus H. Troberg

ABSTRACT

The purpose of this paper is to analyze the Finnish accounting rules currently in effect, the rules proposed in the *Draft to New Accounting Act* of 1990, and to compare the two sets of rules to one another as well as to international accounting standards and principles. The analysis and comparison include the objectives of financial reporting, valuation and measurement principles and concepts, specifics of certain rules (standards), and the formats of financial statements.

The accrual principle of accounting is the guiding principle for revenue and expense recognition in Finland. The objective of financial reporting is defined as to determine distributable profits. The philosophical stand of the Finnish accounting school differs from those of other nationalities in that not only dividends but also interest and taxes are considered to be distribution of profits. The New Accounting Act will not change this philosophy. Finnish accounting is characterized by strong ties between tax reporting and accounting reporting. As a consequence, Finnish companies tend to pay more attention to tax considerations than to so-called economic reality in preparing their financial reports.

Advances in International Accounting,
Volume 5, pages 25-45.
Copyright © 1992 by JAI Press Inc.
All rights of reproduction in any form reserved.
ISBN: 1-55938-415-8

Most of the changes proposed in the *Draft* were expected because they make Finnish accounting better correspond to international reporting standards. Some questionable practices, however, remain. The rules for determining the cost of inventory, focusing on direct production costs only, differ from those of other countries. The recording of leases and the recording of capital gains are other examples of questionable practices.

INTRODUCTION

The international accounting community has made efforts to harmonize accounting standards and practices among countries. In 1976 the Organization for Economic Cooperation and Development (OECD) issued its Declaration on Investment in Multinational Enterprises. The Annex to this declaration contained financial reporting guidelines for multinational enterprises. In 1978 an OECD committee (Committee on International Investment and Multinational Enterprises) established an ad hoc Working Group on Accounting Standards (permanent as of 1979) to review and support private and governmental efforts underway in member countries and international organizations to improve the comparability of accounting standards (OECD 1976, 14-15).

A study group appointed by the United Nations' Secretary General noted in its report from 1974 a lack of both financial and nonfinancial information in usable form about the activities of transnational corporations. As a result a Group of Experts on International Standards of Accounting and Reporting was formed. This group issued the report *International Standards of Accounting and Reporting for Transnational Corporations* in 1977. In 1979 the United Nations founded the Intergovernmental Working Group of Experts on International Standards of Accounting and Reporting with observers, among others, from the International Accounting Standards Committee (IASC) and the International Federation of Accountants (IFAC).

The IASC, founded by professional accounting organizations in 1972, promotes the worldwide harmonization and improvement of accounting principles used in the preparation of financial statements for the benefit of the public (IASC 1982b). IFAC was established in 1977 with the same membership as the IASC, and is mainly working on the setting of international auditing standards. In 1981 IFAC and IASC agreed that IASC has full and complete autonomy in the setting of international accounting standards. IASC has accordingly been influencing the international accounting community through its issue of International Accounting Standards (IAS). Within Europe the European Community (EC) is the major harmonizing force. Through its Fourth (1978) and Seventh Directives (1983) it has tried to make accounting reporting within the member states more similar to one another.

Finland has been more a follower than an innovator in accounting. The accounting rules and principles come principally in the form of laws. The laws governing accounting in Finland are the Company Act of 1978, the Accounting Act of 1973, and the Business Tax Act of 1968 including amendments. These acts are complemented by the pronouncements of a permanent Accounting Committee (founded to clarify and complement the Accounting Act) and some recommendations of the Association of Certified Public Accountants. Because the accounting rules are in the form of laws, legal and political authorities, in addition to accountants, have significantly influenced the formation of these rules and consequently their content. Through the Business Tax Act, the Finnish state (tax authorities) has had a major impact on accounting practice. As the financing structure of Finnish companies is by international standards highly leveraged, the role of creditors (banks) in the development of accounting reporting rules has by no means been a minor one. The accounting profession in Finland is relatively young. By Anglo-Saxon standards the profession cannot be classified as a strong one, and its role in the standard setting process has not been at all as dominant as its counterparts' roles in the United States and the United Kingdom.

The pressure for change of the current accounting rules and practices comes from the internationalization of the Finnish business community and the internationalization of the financial markets. In order to be able to satisfy the needs of their international interest groups and to raise funds from international capital markets, a necessity considering the high interest rates in Finland in the last few years, Finnish companies have had no choice but to prepare financial reports according to international requirements in addition to reports fulfilling the Finnish legal requirements. In preparing international financial reports, Finnish companies have either been following the U.S. Generally Accepted Accounting Principles (GAAP) or the International Accounting Standards of IASC. Many small- and medium-sized Finnish companies are not able or cannot afford to prepare two separate sets of external financial reports. Even for large companies the dual preparatory work amounts to an extra burden.

The pressure for change of the Finnish accounting rules resulted in the formation of an ad hoc accounting committee (hereafter the Committee) in 1989, the task of which was to prepare a draft for the renewal of the Accounting Act (hereafter the *Draft*) currently in effect. The *Draft* was published at the end of 1990. In choosing between various international accounting alternatives, the Committee selected the EC Accounting Directives as the primary guides in preparing the Draft. The reasoning behind the choice was that, irrespective of whether Finland is joining the EC or not, Finnish companies' international operations predominantly take place within Europe and the EC, implying that the application of EC standards is of greater importance to Finnish companies than the application of other international standards.

The purpose of this paper is to analyze the Finnish accounting rules currently in effect and the rules proposed in the *Draft*, and to compare the two sets of rules to one another as well as to international accounting standards and principles. The analysis and comparison include the objectives of financial reporting, valuation and measurement principles and concepts, specifics of certain rules (standards), and the formats of financial statements. There is no intention to extensively analyze the similarities and differences per se between the various international standards (EC, IASC, U.S. GAAP).

OBJECTIVES OF FINANCIAL REPORTING

The *Draft to New Accounting Act* does not bring about any changes with respect to the stated objective of financial reporting in Finland. The primary objective of financial reporting is to determine the net income (result) of the accounting period and owners' equity, specifically the amount of owners' equity which could be distributed (retained earnings). This implies that the net income of the period is measured for the purpose of determining distributable profits (*Draft* 1990, 91). This position can be criticized for focusing predominantly on the legal determination of distributable profits. In reality, a company's ability to distribute profits is not only dependent on past earnings but also on expected future earnings and planned capital expenditures, as well as its liquidity.

According to the Finnish interpretation of the "cost-income" (the term used for the accrual principle of accounting) theory, distribution of profits includes not only dividends but also interest and taxes. This school of thought is clearly different from international norms, which define interest and taxes as expenses. Interest and taxes are classified also in the Finnish income statements as expenses, but the position of Finnish accounting theory is to view interest and taxes as profit distribution.

In comparing the Finnish objectives of financial reporting to the American, the question emerges whether defining the objectives of financial reporting should concurrently be tied in with defining the users of financial reporting. The Finnish accounting school does not define the users. However, regarding not only dividends but also interest and taxes as part of profit distribution, can be interpreted to imply that financial reports are not only directed to owners but also to creditors, governmental administrators and external users in general. In the United States, the *Statement of Financial Accounting Concepts No. 1* (FASB 1978) defines the objectives of financial reporting as to provide information useful for making economic decisions. This is defined more closely as information useful to investors and creditors and other users in making rational investment, credit, and similar decisions and for predicting, comparing, and evaluating potential cash flows to them in terms of amount, timing, and related uncertainty. The influence of ownership and investor interest on financial reporting has in the Anglo-Saxon countries been clearly stronger than in other countries. An example of this is the mandatory disclosure of earnings per share, only recently introduced in Finland and affecting only companies listed on the Helsinki Stock Exchange. The importance of stock exchanges and in that respect owners' equity as a form of financing is greater in the Anglo-Saxon countries than in other countries and consequently explains the strong emphasis on ownership and investor interest in the former.

The EC's Fourth Directive (1978, article 2, para. 3) defines the objective of financial statements as to give a "true and fair view" of a company's assets, liabilities, financial position, and profit or loss. No specific interest group is singled out in the Directive but the statements are directed to the public in general. This corresponds also to the IASC position. According to IASC, financial reporting should provide information for making economic decisions. This is described as information about the financial position, results of operations and the change in the financial position of a company (IASC 1989).

ACCOUNTING PRINCIPLES

The EC Accounting Directives were used as guidelines in preparing the *Draft*. As a result the "true and fair" convention is proposed to be introduced in the New Accounting Act. The convention is not a part of the current Accounting Act from 1973 but it has still been thought of as a guiding principle. Because practice has shown that some companies do not fully adhere to this principle, it was considered to be better to spell it out in the New Acounting Act. The content of this convention is not defined in the *Draft*, but it is thought that accounting theory and practice will continously shape it.

The issues of asset valuation and income determination (measurement) have proven most controversial. Many attempts have been made to settle them, among others, *Dynamic Balance Sheet* (*Dynamische Bilanz*, Eugen Schmalenbach 1919), *The Theory and Measurement of Business Income* (Edgar Edwards and Philip Bell 1961), *A Tentative Set of Broad Accounting Principles for Business Enterprises* (Accounting Research Study No. 3, Robert T. Sprouse and Maurice Moonitz 1962), *Accounting, Evaluation and Economic Behavior* (R.J. Chambers 1965), *Theory of the Measurement of Enterprise Income* (Robert Sterling 1970), *Theory of Accounting Measurement* (Yuji Ijiri 1975), *Tell It Like It Was* (Robert N. Anthony 1983), and *Cash Flow Accounting* (T.A. Lee 1984). There is no intention to discuss the different schools of thought relating to this matter in this paper but to present the position of Finnish accounting.

Finnish accounting adheres to the principle of entry value accounting, more specifically acquisition (historical) cost accounting. That has been reinforced by the strong link between financial reporting and taxation in Finland. The tax authorities' position on verifiability is very strict. Deviation from the main principle of acquisition cost accounting comes in the form of a possible revaluation of certain assets and the application of the lower of cost or market principle.

Revaluation of long-term assets, generally land and buildings, is allowed to some extent. The exit value, specfied as the probable sales price (realizable value), constitutes the upper monetary limit for the revaluation of an asset. Revaluation is not allowed on a regular basis but is viewed more as an exceptional measure. In the *Draft*, the exceptional feature of a revaluation is underlined by the specific requirement that the expenses and potential taxes relating to the possible sale of the long-term asset in question have to be taken into account, that is, the net realizable value should be applied. The criticism against the use of exit values is that in most cases the assets in question are not and will not be sold. Technically a revaluation is carried out by writing up the asset in question and recording an amount corresponding to the write-up as a revaluation reserve in restricted owners' equity. The revaluation reserve can be used for an unpaid share issue. Any possible depreciation relating to the revaluation increment of an asset is not tax deductible. In many cases, revaluations have been used for "window-dressing" reasons, that is, to increase the amount of total assets and, above all, to improve the company's debt-to-equity ratio.

Current entry values are generally not used in Finnish financial reporting. An inflation accounting standard similar to the *Statement of Standard Accounting Practice No. 16* in the United Kingdom or the *Financial Accounting Standard No. 33* in the United States or their revisions has not seriously been considered in Finland despite that inflation at times has been running at fairly high rates. A fairly common practice among large international companies, to disclose current cost of fixed assets and current cost depreciation is rare among Finnish companies. This is in spite of the fact that large Finnish companies commonly use current values in their internal reporting.

The internationally common lower of cost or market principle for current assets is also applicable to Finnish accounting. Market is specified in the Finnish Business Tax Act (1968, $28) as either current entry value or exit value. There have, however, been cases, which show that the principle of cost recovery has not strictly been followed in practice. Because of some major controversies with respect to this matter during the last few years, the *Draft* contains a specific requirement according to which costs that are not likely to be recovered must be written off. Similarly, probable future losses cannot be amortized but have to be written off (*Draft* 1990, 122-123). These requirements are congruent with the corresponding EC and IASC rules.

The accrual principle of accounting is the guiding principle for revenue and expense recognition in Finland. Use of the percentage-of-completion method, which internationally is well accepted in the accounting for large long-term projects, has not been permitted in Finland. The strict adherence to the accrual principle has been criticized for leading to the distortion of the financial reports of companies operating in the building and construction industry. The New Accounting Act will, if approved as proposed in the *Draft*, remedy this by allowing the use of the percentage-of-completion method.

Internationally accepted accounting principles and conventions applicable to Finnish accounting in addition to the above are going concern, conservatism, materiality, cost-benefit, substance over form, completeness, comparability, and timeliness. All of these principles have not been strictly followed in practice. There are for instance cases in Finnish accounting where form has taken priority over substance. This is the consequence of the fact that most of the accounting rules come in the form of detailed legal rules. If these rules stipulate a certain way of reporting which happens to present a company in a more favorable light than the substance of the situation would require, the tendency has many times been to let the legal requirement override the substance requirement. Whether the "true and fair" convention will change this remains to be seen. Both the Accounting Act (1973) currently in effect and the proposal in the *Draft to New Accounting Act* require the financial statements to be made public within six months from the end of the accounting period. Whether that satisfies the timeliness criteria can be questioned. Most of the companies noted on the Helsinki Stock Exchange make their financial statements public within three to four months.

THE INFLUENCE OF TAXATION

Finland is different from the United States, the United Kingdom and the Netherlands in that the ties between taxation and financial reporting are very strong. This means that certain major expense items have to be fully incorporated in the official financial statements in order to be tax deductible. As a consequence, in preparing financial statements "economic reality" considerations are many times secondary to tax and in that respect cash flow considerations. Financial reporting in Finland has been heavily criticized for providing "tax minimization" statements rather than "economic reality" statements. The major differences between "tax reality" and "economic reality" emerge from the handling of depreciation and the use of certain reserves.

The Finnish Business Tax Act (1968) provides depreciation rates for depreciable assets. Until the beginning of the 1980s, these rates were the only ones used in external financial reporting. These rates are not based on economic reality but give companies the possibility to use higher depreciation rates than the economic useful life of an asset and the corresponding consumption thereof would warrant. The major distortion, however, stems from the fact that companies can apply any depreciation rate between 0 and 30% to machinery and equipment (Business Tax Act 1968, $ 30). This flexibility means that Finnish companies use the maximum depreciation rate or a rate close to the maximum in profitable years, while depreciating hardly anything in loss years. This is a major violation of the consistency principle and contrary to the EC requirements, according to which depreciation has to be carried out irrespective of the profitability of the accounting period (Fourth Directive 1978, article 35, para. 1). Largely due to the internationalization of Finnish firms, the practices of most large companies have improved in the sense that in the income statement they present depreciation according to plan (based on estimated useful life) as the "real" depreciation. The difference between planned depreciation and tax depreciation is disclosed separately in the income statement. The New Accounting Act would make it mandatory for large companies to apply depreciation according to plan. Small companies, however, are not bound by this requirement.

For a long time the creation of reserves for tax deduction purposes has been a permissable practice in the Nordic countries. The most important Finnish reserves are the inventory reserve, the operating reserve, and the investment reserve. The inventory reserve was originally created as a protection against inflation. The first-in first-out method is the inventory method required in the Business Tax Act. This leads to high taxable income in times of inflation if there are no other means of reducing taxable income. The use of the last-in first-out inventory method is not permitted. Lately, also the weighted-average method has entered into practice. A company's inventory reserve can constitute 0-25% (earlier 0-50%) of the acquisition cost of the inventory remaining at the end of the accounting period. This means that up to 25% of the cost of the inventory can be expensed although it is not sold. This is a major violation of the accrual principle of accounting.

Technically the inventory reserve system works so that the allocation to the reserve is recorded as an expense in the income statement with an equal amount being added to the reserve itself in the balance sheet. The inventory reserve is presented between liabilities and owners' equity. It is always the change in the reserve that affects the income statement while the accumulated amount of the allocations constitutes the reserve in the balance sheet. In practice, the reserve is used for smoothing net income and minimizing taxes, that is, allocations (expense) to the reserve are made during profitable years while the reserve or a part of the reserve is dissolved (revenue) in loss-bringing years. Because large companies are obliged to openly present the allocation to or dissolution of the reserve under appropriate headings, including income before the appropriation to (dissolution of) the reserve, external readers receive today a clear picture of the impact of the inventory reserve system on the financial statements. Small companies (earlier also large companies) hide the effect of the reserve system by including the allocation (dissolution) in cost of sales and deducting the accumulated reserve amount directly from the inventory account.

The operating reserve can constitute 0-30% of the salaries and wages of the accounting period, which are subject to income taxes. Technically the operating reserve system works the same way as the inventory reserve system. The creation of an operating reserve system gave service companies an opportunity to more easily smooth their net income. A manufacturing company can and generally does use the inventory reserve system and the operating reserve system at the same time, but the systems are tied to each other in the respect that both reserves cannot simultaneously be at their maximum levels.

The investment reserve system allows the creation of a reserve of 0-20% of the net income before taxes. The investment reserve is primarily intended to be used for capital expenditures but also to cover costs of research and development and employee training. The allocation to the reserve is recorded as an expense while the dissolution reduces the depreciable amount of the property, plant, and equipment acquired by the amount of the dissolution.

Overall, companies are allowed total flexibility within the stipulated rate limits to allocate to or to dissolve their inventory reserves and operating reserves as well as to allocate to their investment reserves. When this is coupled with the flexible depreciation rates of machinery and equipment, it is not surprising that financial reporting in Finland has been criticized for providing distorted financial information. In preparing financial statements, tax considerations are given priority over "economic reality" considerations. It has to be pointed out that the reserve amounts and depreciation amounts by which a company "arranges" its reported net income are by no means minor amounts. With respect to large companies, the situation has improved as they are required to clearly present under proper headings in the financial statements the amounts of the reserve allocations/dissolutions of the accounting period, the current balances of the reserves, and the differences between planned and tax depreciation. Because the upper limits of the reserves are tax political decisions they may change from year to year and they have been declining over the last few years.

In developing the *Draft to New Accounting Act*, the Committee considered the problem of the close ties between financial reporting and taxation. The Committee selected not to cut these ties. Instead the Committee proposes, similarly to the French model, the possibility (but not requirement) to eliminate the reserve allocations (dissolutions) and depreciation differences from the consolidated financial statements. The rationale behind this is that it is the legal entities and not the consolidated entities which are taxed. Furthermore, consolidated statements are internationally regarded as more important than the statements of the individual legal entities. In the preparation of the consolidated financial statements, the amounts eliminated would be split between net income and consequently owners' equity and deferred taxes. From a financial analysis point of view, the question is whether or to what extent deferred taxes will be paid. The problem is similar to the deferred taxation problem in the United Kingdom and the United States. As long as a company's level of operations remains the same or is rising, it is unlikely that the deferred taxes will have to be paid in the foreseeable future. Whether these amounts go under the heading "deferred taxes" or "reserves" seems irrelevant from a financial analysis perspective. The Committee did not reach unanimity with respect to the treatment of the reserve allocations (dissolutions), reserves, and depreciation differences in the consolidated financial statements.

SOME SPECIAL FEATURES

Inventory

In addition to the reserve features, Finnish financial statements differ substantially from those of most other countries with respect to the cost structure of inventory and in that respect cost of goods sold. Before clarifying the cost structure, it is necessary to briefly describe the conceptual confusion that exists in the Finnish accounting terminology.

The conceptual problem arises from the use of the terms variable, fixed, direct, and indirect costs. The Business Tax Act of 1968 ($ 14) defines the acquisition cost of inventory as being the direct costs. The Accounting Act of 1973 ($ 13) defines the acquisition cost of inventory as being the variable costs. In other words, the concept pairs of variable-fixed and direct-indirect have been used interchangeably as if they were synonyms. The *Draft* does not bring much new light on this issue. In the discussion of the acquistion cost of an asset, variable cost is defined correctly in the *Draft* as varying in relation to production volume, that is, as a behavioral element. Fixed costs are not defined in the *Draft* but the production related fixed costs are described as meaning indirect material costs, the wages of the production management, the depreciation of property, plant and equipment used in the production, and the administrative costs of the plant. This may give the impression that fixed cost by definition is a larger concept than indirect cost. In the discussion of the *Draft*, a comparison is also made to the EC Fourth Directive (1978, article 35, para. 3), in which the concepts direct and indirect are used. Overall, the concepts variable, fixed, direct, and indirect costs are not clearly defined in the Draft. Whether this is purposely done or due to ignorance is difficult to tell. In this context, it is necessary to draw attention to a certain feature of the Finnish income statement, namely its division into two major expense categories, variable and fixed. This may explain the adherence to the variable-fixed cost concept instead of to the direct-indirect cost concept in defining the acquisition cost of an asset.

In describing the cost structure problem, the author takes the liberty of interpreting the use of the term variable costs in Finnish accounting as generally meaning direct costs. In Finnish accounting, the cost of inventory (and assets in general) in a manufacturing company is determined as consisting of the direct costs, that is, direct material and supplies and direct production labor. The internationally common practice to also include production overhead in the inventory cost is not allowed in Finland. The problem that the current Finnish practice causes with respect to international comparability is discussed in the *Draft*. The potential consequences that excluding production overhead from the inventory cost may have for liquidity evaluation (current ratio) of Finnish firms in an international context are not discussed in the *Draft*. The relative difference in the cost structure of inventory between Finnish companies and companies of other national origin has widened during the last few years because of the increased use of capital intensive production technology in the manufacturing process.

The New Accounting Act would not change the main rule for determining the cost of inventory, that is, the cost of inventory is defined as the variable ("direct") costs of the good. However, if the fixed ("indirect") production related costs are significant and when the size of a company's operations would so justify, the company would be allowed to include fixed production costs in its inventory value. Such a cost structure would better correspond to international practice including that of most EC countries. As has been stated, the fixed production costs are in the *Draft* specified as primarily meaning the costs of indirect material and supplies, the wages of the production management, the depreciation of the manufacturing plant, machinery and equipment, and the administrative costs of the plant. The possibility to incorporate fixed production costs into the inventory costs and consequently depart from the rule is mainly intended for large companies. The Committee argued that it would be technically difficult for small companies to include fixed production costs in their inventory costs, practically meaningless for determining their net income and, furthermore, most of them would not find it useful to change their practice (*Draft* 1990, 103). The author has to disagree with the Committee on this issue. The increased use of capital-intensive production technology means that production overhead has increased both absolutely and proportionally, while the direct labor content has declined. In the long run, neither large nor small companies can afford to ignore the size and importance of the so-called production overhead in their asset valuation, internal product profitability, and price setting calculations. Some of this information could also be used externally.

Leasing

No difference is made between operating type leases and capital (finance) type leases in Finnish accounting. Irrespective of the nature of the lease contract, all leases are treated as operating leases, that is, the lease payment(s) is treated as an expense of the accounting period. Some information about long-term capital-type leases has to be disclosed in the notes but capitalization of leases is neither required nor allowed. The *Draft* does not bring any changes to existing practice. This is somewhat surprising because most of the accounting changes proposed in the *Draft* are for the purpose of making Finnish accounting practice more comparable to international accounting. Whether there is an intention to change the current leasing practices through some other legislation or authoritative recommendation is unknown.

Commercial Paper

The acquisition cost of securities and commercial paper has to be written down if market value has significantly declined and the decline is of a permanent nature. This accounting standard (rule) is common in many countries including Finland. The annual reports for 1990 of some major Finnish companies, however, show that there are Finnish companies, which, with the consent of the auditors, do not find the rule binding. In one instance, the market value of securities (shares in an ailing company) had been significantly below the acquisition cost for more than a year. The reason for the questionable practice stems from the fact that Finland is currently experiencing a recession. Companies are reporting sufficiently bad results without having to incorporate unrealized losses on securities.

A unique feature of Finnish accounting is that once the acquisition cost of a security has been written down, it may not be written up again. The internationally permissable practice to make a write-up (at most to the original acquisition cost) of a security, if the market value has significantly risen and the rise is of long-term nature, is not applicable to Finnish accounting.

Foreign Receivables and Other Debts

The treatment of accounts receivable and other debts denominated in foreign currencies varies due to the flexibility allowed by and inconsistencies contained in the current Finnish regulations. The Accounting Act of 1973 ($ 15) states that a receivable denominated in a foreign currency can "at maximum" be recorded at the closing exchange rate of the accounting period . This means that if the exchange rate from the company's perspective has improved as compared to the transaction date, the company can, if it so chooses, retain the rate of the transaction date and consequently avoid to record any gains. If the rate has moved in an unfavorable direction, the closing rate has to be used and consequently an exchange rate loss has to be recorded. This type of practice is in line with the conservatism principle, yet produces inconsistent results.

According to the rules of the Accounting Act (§ 15), payables denominated in a foreign currency can either be recorded at the transaction date rate or the closing rate. This means that a company can choose whether or not to record an exchange rate loss. Similarly it can choose whether or not to record an exchange rate gain. This is different from the ruling concerning foreign receivables and cannot be said to be in line with the principle of conservatism. Furthermore, exchange rate losses on long-term debt have in many instances been capitalized and expensed first in later accounting periods.

The EC Directives do not contain any specific ruling on the treatment of receivables and debts denominated in foreign currencies. The position of the International Accounting Standards Committee is that receivables and debts including long-term debts have to be recorded at the closing rate (IAS 1982a). In the United States, the *Financial Accounting Standard (FAS) 52* (FASB 1981) contains a similar requirement. The IAS No. 21 application means that exchange rate gains and losses (disregarding any effects of hedging activities) have to be recorded and consequently affect reported net income. This is not necessarily a satisfactory solution. Many of the recorded exchange rate gains and losses will never materialize, at least not to their full extent, because of the possible reversal of the exchange rate movements before the settlement date. The proposal in the *Draft* is that the closing rate should be used for both receivables and debts denominated in foreign currencies. This would make the Finnish regulation congruent with the IASC and FASB regulation.

Depreciation

The current Finnish accounting practice allows but does not require companies to use depreciation according to plan. Depreciation according to plan means that a systematic depreciation schedule based on the economic useful life of the asset is applied. In practice the Business Tax Act (1968) has been the main influencing force on the depreciation behavior of Finnish companies. The Act allows, as has been related earlier, major flexibility with respect to applicable depreciation rates. The economic life of the asset is irrelevant in determining these rates. In preparing financial reports, small- and medium-sized companies pay attention almost exclusively to tax considerations and consequently present only so-called tax depreciation in their financial statements. The companies quoted on the Helsinki Stock Exchange present both depreciation according to plan and "tax" depreciation. In both the Fourth EC Directive (1978, article 35, para. 1) and the IAS No. 4 (1976) depreciation according to plan is demanded. International comparability of Finnish companies would consequently be enhanced by the position taken in the *Draft*, that is, the new regulation would make depreciation according to plan mandatory. Only small companies would be exempted from this requirement.

Research and Development

Contrary to the U.S. practice, research and development costs can be capitalized in Finland. The Business Tax Act governs the actual behavior. The research and development costs can either immediately be expensed or amortized over a period of two to ten years (Business Tax Act 1968, § 25). In profitable years companies tend to expense all their research and development costs while capitalization is typical for unprofitable years. The possibilites of capitalization under the IASC rules are relatively restricted.

CONSOLIDATION

The main consolidation rules are incorporated in the Company Act of 1978. The intention of the Committee is to make the consolidation rules a part of the New Accounting Act. According to the current rules (Company Act 1978, chapter 1, para. 2), the financial statements of two companies have to be consolidated when one of the companies (parent) owns more than 50% of the voting shares of the other (subsidiary). In modifying and changing the current regulation, the Committee used the Seventh EC Directive as the guideline. In accordance with the EC rules, the Committee proposes that the consolidation of financial statements is necessary not only when a (parent) company controls the majority of the voting shares of another company (subsidiary) but also when a company exerts controlling influence on another company by being able to appoint more than half of the members of the administrative, managerial, or supervisory body of the other company (Seventh Directive 1983, article 2, para. 2).

The purchase method is the consolidation technique currently in use in Finland. The *Draft* brings no changes to this. Goodwill has been amortized over a maximum period of ten years. Within the EC the maximum period for goodwill amortization is five years (Fourth Directive 1978, article 37, para. 2), and a direct write-off against shareholders' equity is also possible. Such a practice is common in the United Kingdom and the Netherlands. According to the *Draft*, unamortized goodwill should be disclosed separately in the consolidated balance sheet below property, plant and equipment, and other long-term investments. Unamortized negative goodwill should be shown separately in the balance sheet between liabilities and shareholders' equity (after the reserves). Amortization of goodwill and negative goodwill should be presented separately in the income statement below depreciation of property, plant and equipment. The minority interest's share of the net income has to be disclosed separately. Similarly, the minority interest has to be presented separately in the consolidated balance sheet between liabilities and shareholders' equity (*Draft* 1990, 158, 170).

Investments in other companies are reported under the cost method, irrespective of the proportion of the ownership. The New Accounting Act will change this in accordance with international practice, that is, 20-50% ownership (affiliated/ associated companies) will be accounted for according to the equity method.

QUESTIONABLE PRACTICES

There are practices in Finnish accounting that violate the all-inclusive-income statement concept. Some Finnish companies deduct a part of their taxes directly from shareholders' equity, thus bypassing the income statement. Originally only taxes stemming from preceding accounting periods were recorded in this way. Some companies, however, extended the practice to also include the taxes, or some part of the taxes, of the current accounting period. Many companies add capital gains on the sale of property, plant and equipment directly to shareholders' equity. Such a practice is rather typical in profitable years, while the tendency in unprofitable years is to include capital gains in the income statement. The described tax and capital gains practices are most questionable and cannot be deemed as internationally acceptable. The frequency of these practices has clearly declined during the last few years.

So-called "valuation items" are used in the balance sheet to incorporate certain transaction related developments, which by the end of the accounting period are not yet finalized and the outcome of which is therefore pending. Examples are exchange rate effects on foreign receivables and other debts which are not settled by the end of the accounting period. Other changes in outstanding receivables and debts can be included in the valuation items. Incorporating unrealized exchange rate change effects on foreign receivables and debts into the valuation items, thus bypassing the income statement, may be regarded as an acceptable alternative to the recording of these in the income statement, because a valuation adjustment is also used in the United States. Revaluations of property, plant and equipment can also be included in the valuation items.

Finnish accounting practice differs from general international practice in that the costs of pension obligations incurred in a certain accounting period need not be expensed but can be capitalized as valuation items. International practice is to expense pension costs in the period during which the obligations have arisen and to which the costs therefore belong. The New Accounting Act would, if approved in its present form, require Finnish companies to expense pension costs in the accounting period during which the obligations have been incurred. Hence Finnish accounting practice would become similar to international practice in this respect.

Interest on construction-related loans can be capitalized. In this respect the Finnish accounting practice is similar to the U.S. accounting practice. The interest is, however, not capitalized as part of the acqusition cost of the construction in question but is capitalized separately as a valuation item.

FORMATS OF FINANCIAL STATEMENTS

Compared with the current situation, the New Accounting Act would make the formats of Finnish financial statements better correspond to the EC requirements but still retain some differences. The proposed balance sheet format does not, however, substantially differ from that of the Accounting Act of 1973, except for the fact that the order of presentation is reversed. The suggestion in the *Draft* is to start with the less liquid assets as compared to the more liquid assets as determined in the current Act. Also the order of the equities would be reversed and start with shareholders' equity and end with current liabilities (Table 1).

The *Draft* does not bring about any major changes in the format of the income statement as compared to the requirements of the current Accounting Act. The Act of 1973 applies a behavioral classification of the expense items by dividing the expenses (costs) into two major categories, namely variable and fixed. Although some of the headings proposed in the *Draft* are modified from those in the Act of 1973, the variable-fixed classification would still form the basis for the income statement structure (Table 2). The variable expenses consist primarily of the production costs while the fixed expenses consist of the marketing and administrative costs. In reality, the degree of expense variability as measured over one accounting period is difficult to determine. The accuracy of the variable-fixed expense classification can be questioned and differs among companies.

As related earlier, the New Accounting Act would allow companies with significant indirect production related costs (production overhead) to include such costs in the determination of the company's inventory costs, when the size of the company's operations would so justify. Companies which include indirect production costs in their inventory calculations would have to use a functional income statement approach. The format of the income statements of such companies would better correspond to the formats used internationally than the format of the current Finnish accounting practice or the principal format recommended (required) in the *Draft to New Accounting Act*.

Neither in the Accounting Act of 1973 nor in the *Draft* is there a requirement of a statement of changes in the financial position or of cash flows. Most large and many medium-sized Finnish companies have, however, prepared a statement of changes in the financial position for a long time. The formats used have varied but some form of change in working capital has dominated during the last few years.

Table 1. Finnish Balance Sheet as Proposed in the Draft to New Accounting Act

ASSETS

PROPERTY, PLANT AND EQUIPMENT AND
OTHER LONG-TERM EXPENDITURES
  Intangible Assets
    Formation and Restructuring Costs
    Research and Development Costs
    Copyrights, Patents and Other Contractual Rights
    Goodwill
    Other Long-term Expenditures
    Prepayments
  Property, Plant and Equipment
    Land and Water Areas
    Buildings and Constructions
    Machinery and Equipment
    Other Expenditures
    Prepayments and Work-In-Progress
  Long-term Investments
    Shares
    Notes Receivable
    Other

VALUATION ITEMS

INVENTORY AND FINANCIAL ASSETS
  Inventory
    Material and Supplies
    Work-In-Process
    Finished Goods
    Other Inventory
    Prepaid expenses
  Receivables
    Accounts Receivable
    Notes Receivable
    Accrued Receivables
    Other Receivables
  Investments
    Shares
    Other
  Cash On Hand and In Bank

EQUITIES

SHAREHOLDERS' EQUITY
  Paid-In Capital, at par
  Other Shareholders' Equity
  Retained Earnings
  Net Income of the Period

(*continued*)

Table 1.   (Continued)

RESERVES
    Accumulated Depreciation Difference
    Voluntary Reserves
        Investment Reserve
        Inventory Reserve
        Other Reserve
    Mandatory Reserves (to take into account expected future
    losses/expenses materializing from existing obligations)

VALUATION ITEMS

LIABILITIES
    Long-term Liabilities
        Bonds
        Convertible Bonds
        Loans from Financial Institutions
        Pension Loans
        Deferred Revenues
        Accounts Payable
        Other Long-term Debt
    Current Liabilities
        Loans from Financial Institutions
        Pension Loans
        Deferred Revenues
        Accounts Payable
        Accrued Payables
        Other Current Liabilities

DISCLOSURES

The list of disclosure requirements in Finnish accounting is rather long. Segment reporting has not been well developed in Finland because no legal or other requirement has existed in this respect (except for companies quoted on the Helsinki Stock Exchange). The New Act would require the disclosure of sales (turnover) by industry (category of activity) and geographical area. This requirement can be regarded as rather basic. The EC Fourth Directive (1978, article 43, para. 1) demands disclosure of turnover and number of employees by categories of activity and geographical areas. These disclosures are still less than U.S. requirements. According to FAS 14 (FASB 1976), revenues, operating profit, identifiable assets, and potentially also depreciation and capital expenditures should be disclosed by industrial segment. Revenues, operating profit, and identifiable assets should be disclosed by geographical segment. The identity and amount of revenues from customers providing more than 10% of the total revenues of a company should also be disclosed. The

Table 2. Finnish Income Statement as Proposed in
the Draft to New Accounting Act

Sales (net)
Cost of Sales
    Material and Supplies/Goods
    Salaries and Wages
    Social Expenses
    Other Variable Expenses
    Production for Own Use (minus)
    Change in Inventory (plus or minus)
Sales Margin ($\approx$"Gross Margin")
Other Operating Expenses
    Salaries and Wages
    Social Expenses
    Other Expenses
    Other Operating Revenues
Contribution from Operations
Depreciation
Profit/Loss from Operations
Financial Revenues and Expenses
    Dividends Received
    Interest Revenues
    Other Financial Revenues
    Interest Expenses
    Other Financial Expenses
    Adjustments of the Capitalized Acquistion Costs of
    Short-term Investments (plus or minus)
Net Income (Loss) before Occasional Items, Changes in
 Reserves and Taxes
    Occasional Revenues
    Occasional Expenses
Net Income (Loss) before Changes in Reserves and Taxes
Increase/Decrease in Accumulated Depreciation Difference
Increase/Decrease in Voluntary Reserves
Taxes
Net Income of the Period

criticism of segment disclosures universally is the flexiblity that the management of a company has in defining a segment.

The EC rules require that significant after-balance-sheet-date events should be disclosed in the company's annual report (Fourth Directive 1978, article 31, para. 1). The Finnish Company Act of 1978 (chapter 11, para. 9) contains a similar requirement. If a company is not disclosing a significant after-balance-sheet-date event in its annual report, the auditor has to bring attention to this fact in his/her audit report. Recent developments in some troubled Finnish companies have raised the question about the extent to which the described obligations are fulfilled in practice.

SUMMARY

The internationalization of Finnish firms and the internationalization of capital markets have created a pressure on Finnish accounting authorities to internationalize the Finnish accounting rules. The EC Accounting Directives were used as guidelines in preparing the *Draft to New Accounting Act*. The *Draft* does not accommodate all the EC requirements. It was considered that Finnish companies need time to adjust their reporting to the proposed accounting changes. More extensive accounting changes would make the adjustment process difficult. The pressure for revision of the current rules originates also from some major bankruptcy cases in which the financial reporting has been heavily questioned and criticized.

Most of the proposed changes were expected because they would bring Finnish accounting rules more in line with international rules. Surprisingly, the principal rule for determing the cost of inventory remains the same, that is, the cost of inventory is determined to consist of direct production costs only, thus excluding production overhead. The leasing rules would not change which means that leases which internationally would be classified as capital leases will still be treated as operating leases in Finnish accounting. The strong ties between tax reporting and financial reporting will seemingly be retained on the individual legal entity level but are proposed to be eliminated from the consolidated level. How this will work in practice remains to be seen.

The New Accounting Act would become effective as of January 1, 1993. The permanent Accounting Committee (mentioned in the introduction) would advise companies on the application of the New Act and complement the rules by its own pronouncenments.

REFERENCES

Accounting Act 1973. Helsinki: Government Printing Office.

Business Tax Act 1968. Helsinki: Government Printing Office.

Committee on the Renewal of the Accounting Act. 1990. *Draft to New Accounting Act*. Helsinki: Government Printing Office.

Company Act 1978. Helsinki: Government Printing Office.

European Economic Community. 1988. *Fourth Council Directive*. Brussels: EEC.

_____. 1983. *Seventh Council Directive*. Luxembourg: EEC.

Financial Accounting Standards Board. 1976. Statement of Financial Accounting Standards No. 14: *Financial Reporting for Segments of a Business Enterprise*. Stamford, CT: FASB.

_____. 1978. Statements of Financial Accounting Concepts for Business Enterprises No. 1: *Objectives of Financial Reporting by Business Enterprises*. Stamford, CT: FASB.

_____. 1981. Statements of Financial Accounting Standards No. 52: *Foreign Currency Translation*. Stamford, CT: FASB.

International Accounting Standards Committee. 1976. International Accounting Standard No. 4: *Depreciation Accounting*. London: IASC.

_____. 1982a. International Accounting Standard No. 21: *Accounting for the Effects of Changes in Foreign Exchange Rates*. London: IASC.

_____. 1982b. *Objectives and Procedures*. London: IASC.

_____. 1989. *Framework for the Preparation and Presentation of Financial Statements*. London: IASC.

Organization for Economic Cooperation and Development. 1976. *International Investment and Multinational Enterprises*. Paris: OECD.

THE PREDICTION OF INTERNATIONAL ACCOUNTING STANDARDS PROFITS FROM FINANCIAL STATEMENTS OF FINNISH FIRMS

Eero Kasanen, Juha Kinnunen,

and Jyrki Niskanen

ABSTRACT

The main purpose of this paper is to test alternative models for predicting IAS profit figures from financial statements based on (non-IAS) Finnish accounting regulations. Such model(s) should be potentially useful at least for security analysts, firms' accountants as well as for researchers. Finland provides a particularly interesting case for this kind of study because of some important differences between the IAS and Finnish accounting rules. The findings show that a relatively

Advances in International Accounting,
Volume 5, pages 47-73.
Copyright © 1992 by JAI Press Inc.
All rights of reproduction in any form reserved.
ISBN: 1-55938-415-8

simple analytical model based on the indirect method of IAS profit determination tends to outperform statistical regression models based on different estimation techniques. This suggests that when the reported net income figure is adjusted only for a few important differences between the Finnish and IAS accounting rules, the statistical approach can do little in further improving IAS profit predictions.

INTRODUCTION

In the past two decades, the accounting profession has responded on four different levels to the growing demand for internationally comparable accounting information. On the global level, the International Accounting Standards Committee has promulgated by now 31 recommended accounting standards since its foundation in the early 1970s. On the European level, the European Community has issued two mandatory directives (4 and 7) designed to harmonize accounting practices in the member countries. On the individual country level, some non-EC countries (e.g., Finland) have recently responded to the challenge by taking measures to change their accounting legislation to facilitate international comparability of financial statements. Finally, the response on the individual firm level is manifested by the fact that many firms operating on the international markets have voluntarily disclosed dual sets of financial statements—one to comply with domestic accounting requirements and the other to provide useful information for the international financial community.

This paper tackles the response that has taken place on the last mentioned firm level. To be more specific, our aim is to test alternative models that could be used for predicting IAS profits for firms that do not already disclose such profit figures voluntarily.[1] Such prediction models may be useful at least in the following contexts.

1. *Fundamental security analysis.* Corporate net earnings can be regarded as the most important single variable in fundamental security analyses aimed at assessing the future earning power of the firm. If such analyses take place in an international environment, the analyst must use earnings information that is based on uniform accounting methods to make sure that statistics such as EPS or PE ratios are comparable from one country to another. In case the firm voluntarily discloses IAS profit figures, the analyst can of course use them, but if the firm does not, the analyst has to estimate them.

2. *Firm's internal use.* Depending on the differences between national accounting rules and International Accounting Standards, the computation of the precise IAS profit for the dual financial statements is not a job for a layman

because in most cases it will be a difficult and time-consuming task requiring real accounting expertise.[2] Even if modern information technology is available for its precise computation, a shortcut prediction formula for the IAS profit may yet be useful for the preparers of financial statements. At the minimum, it can serve as a benchmark against which the computed precise profit figure can be compared. In case a remarkable difference is found, it may prove worthwhile to provide an explanation for the difference between the actual and expected profit figures.

3. *Research purposes.*   The availability of firm-specific dual earnings data may provide an opportunity to test some interesting research hypotheses that otherwise would be more difficult to tackle. An example is the income smoothing hypothesis for which the literature suggests various external and internal motives.[3] As far as IAS profit can be regarded as a less smoothed earnings measure than domestic earnings (as is the case, e.g., in Finland), its availability provides an opportunity to test some smoothing motives because it allows the measurement of the degree of income smoothing in individual firms. A prediction model for IAS profit would thus serve as a "data-generator" for those firms which do not disclose dual profit figures voluntarily.

Finland provides a particularly useful case for this kind of study for several reasons. First, as the recent IASC (1988) survey reveals, Finnish accounting practice has the lowest conformity with International Accounting Standards among the 54 countries included in the survey.[4] Because of the important differences between Finnish accounting rules and International Accounting Standards (which will be discussed below), Finnish and IAS profits can be expected to differ significantly. Of course, these differences stress the need for a prediction model not only for security analyses, but also for the firm's internal use mentioned above. Second, because Finnish accounting regulations (especially the depreciation and reserve rules) give the firms a lot of managerial discretion on the determination of annual net income, there are sufficient grounds to assume that their earnings series are smoothed relative to IAS earnings. This emphasizes the need for a prediction model for the research purposes discussed above.[5] Finally, because some Finnish firms have begun to voluntarily disclose their IAS profit figures in the 1980s, there are now data available for the estimation of alternative prediction models as well as for tests of their predictive performance.

The results of this study show that a relatively simple analytical model based on the indirect method of IAS profit determination tends to outperform statistical regression models based on different estimation techniques. This suggests that when the reported net income figure is adjusted for a few important differences between the Finnish and IAS accounting rules, the statistical approach can do little in further improving the accuracy of IAS profit predictions for Finnish firms.

The remainder of this paper is organized as follows. The differences between the Finnish accounting rules and International Accounting Standards are discussed in the following section. The prediction model tested in this study are then described. Thereafter, the data and methodology used in the test of the prediction models are presented. Empirical results are then reported, and a summary is presented in the final section.

DIFFERENCES BETWEEN THE FINNISH AND IAS ACCOUNTING RULES

The main differences between Finnish and IAS accounting rules are briefly the following:[6]

1. According to IAS, inventory values include manufacturing overhead. The Finnish rules, on the other hand, require that the inventory values be determined on the basis of direct costs. Thus, the inclusion of overhead in the inventory values is not allowed.

2. The Finnish firms usually apply a geometrically degressive declining balance method of depreciation for fixed long-term assets, which is required by tax rules. The IAS recommend the straight-line method of depreciation during the estimated economic life of the asset.

3. According to IAS, financial lease contracts must be capitalized on the lessee's balance sheet. This is not allowed according to the Finnish regulations.

4. According to IAS, 20%-50% owned affiliates should be valued according to the equity method of accounting. In Finland, the cost method is applied for such affiliates.

5. In Finland, the unfunded pension obligations are not treated as balance sheet debt, and the changes in that debt do not affect income. The IAS require their explicit inclusion in the financial statements.

6. The Finnish rules allow firms to create various untaxed reserves. Because increases (decreases) of these reserves are debited (credited) in the income statement, they have direct effect on the reported Finnish earnings number.[7] Furthermore, Finnish regulations allow firms to debit income taxes against shareholders' equity (retained earnings), which is not allowed by the IAS.

7. According to IAS, companies may use either "the percentage of completion" or "the completed contract method" of accounting for long-term construction projects. In Finland, only the latter method is normally allowed. The IAS require that even if "the completed contract

method" is used, the foreseeable losses have to be reported during the project, which is not permitted in Finland.

Given these differences between the accounting rules, it can be expected that IAS profits tend to be larger than Finnish profits. The main reasons for such expectation arc the following.

1. Assuming a constant growth rate for capital investments, Finnish accelerated depreciation should generally be larger than the straight-line depreciation of IAS when the firms run out of other types of tax deductions, that is, when their operations are profitable. In the long run, only those firms that fulfill this condition will survive. Additionally, the depreciation can be done irrespective of when during the fiscal period the asset was acquired. The IAS straight-line depreciation, on the other hand, depends on usage time of the asset, that is, if it was acquired at the end of the fiscal period, it may not be depreciated. Therefore, the timing of the tax depreciation may be "heavier in the front" for this reason.
2. The changes in untaxed reserves do not affect IAS profits. Under periods of inflation and real corporate growth, the untaxed reserves tend to grow from year to year and thus have a decreasing impact on Finnish profits.
3. The share of 20%-50% owned affiliates' earnings affecting IAS profits should usually be larger than the effect of received dividends on Finnish profits.

However, there are some differences which have an opposite impact. For example, the Finnish rules do not allow the depreciation of asset revaluations, while the IAS require their depreciation. Moreover, if the firm determines its income taxes on the basis of (higher) IAS profit, tax expense would be larger and the difference between the Finnish and IAS profits would thus be mitigated somewhat. Additionally, the amortization of capital leases according to the IAS can be expected to decrease net earnings in the early years of the lease when the combined effect of the amortization expense and the interest expense included in the lease payments exceeds the lease payments. Despite these counter-effects, it can be assumed that, in terms of absolute amounts, they are outweighed by the income increasing effects listed above so that on balance most Finnish firms would show larger profits when reporting their results according to IAS than according to Finnish accounting rules.

PREDICTION MODELS

In principle, there are two basic methodological approaches for the determination of IAS net income for any Finnish firms.[8]

1. In the *direct method*, the earnings figure is recalculated from the same basic financial accounting information as the Finnish net income. However, instead of using the Finnish accounting rules for the recognition, valuation, and allocation of revenues and expenditures, IAS are applied.

2. In the *indirect method*, the Finnish net income is adjusted for the difference between Finnish and IAS accounting rules, thus producing the IAS profit as a result.

Of course, these methods generate identical results if all financial accounting information is available, that is, if one has free access to the detailed bookkeeping records of the firm. Apart from the firm's accountants for whom this information is available, neither security analysts nor researchers have such access, and therefore they can use the two methods only to approximate IAS profits. Consequently, the prediction models examined in this study were derived from the two basic approaches under the restrictions imposed by information availability.

Altogether 14 different models were tested in this study. In addition to their underlying methodological approaches, the models differ from each other with respect to their mathematical nature and statistical estimation method, as summarized in Table 1.

Prediction Models Based on the Indirect Method

Given the main differences between the Finnish and IAS accounting rules discussed in the previous section, the general adjustment formula of the indirect method that could be applied if all relevant information were available, is the following.

Component #	
1	Finnish Net Profit (FINPROF$_{it}$)
2	± Net Change in Untaxed Reserves
3	− Provision for Construction Project Losses
4	− Taxes Deducted from Shareholders' Equity
5	+ Income Directly Added to Shareholders' Equity
6	± Share of Affiliates' Net Income (Net Loss)
7	± Unrealized Exchange Gains (Losses)
8	− Dividends from Affiliated Companies (20%-50% owned)
9	− Interest on Capitalized Lease Obligations
10	− Straight-Line Depreciation of the Revaluations of Fixed Assets

11	− Straight-Line Amortization of Capitalized Leases
12	− Increase in Unfunded Pension Obligations
13	+ Lease Expense
14	± Adjustment for Manufacturing of Overhead
15	± Adjustment for the Difference between Declining Balance and Straight-Line Depreciation of Fixed Assets

$$= \text{IAS profit (IASPROF}_{it}) \tag{1}$$

The first model was based on the naive assumption that the net effect of all adjustments 2 through 15 in Equation (1) is zero. Although this assumption contradicts our expectation that the net effect is positive, we nevertheless considered this model because it provided a useful benchmark for the predictive ability of other models.

Model 1

$$E(\text{IASPROF}_{it}) = \text{FINPROF}_{it} \tag{2}$$

where

$E(\text{IASPROF}_{it})$ = Expected IAS profit for firm i in year t
FINPROF_{it} = Reported Finnish profit for firm i in year t

(Subscripts i and t will denote firm and year, respectively, in all subsequent expressions.)

For the second model, four different components of the general Equation (1) were taken into account: #2 (net change in untaxed reserves), #4 (taxes deducted from shareholders' equity), #14 (adjustment for manufacturing overhead), and #15 (adjustment for depreciation). Net change in untaxed reserves and taxes debited against shareholders' equity are, in general, readily available from financial statement information, while adjustments for overhead and depreciation require estimation. The methods used to estimate these two components are presented in the Appendix.

Model 2

$$E(\text{IASPROF}_{it}) = \text{IASCALC}_{it} \tag{3}$$

where

IASCALC_{it} = FINPROF_{it} ± Net Change in Untaxes Reserves − Taxes Deducted from Shareholders' Equity ± Adjustment for Overhead ± Adjustment for Depreciation

Table 1. Summary of the Prediction Models

Model	$E(IASPROF_{it})$	Methodological Background	Mathematical Nature Method	Estimation
1	FINPROF$_{it}$	Indirect	Analytical (Naive)	—
2	IASCALC$_{it}$	Indirect	Analytical	—
3	$b_0 + b_1$ FINPROF$_{it}$ + ...	Indirect	Statistical	Full, OLS
4	STEPWISE (3)	Indirect	Statistical	Stepwise, OLS
5	$b_0 + b_1$ IASCALC$_{it}$ + ...	Indirect	Statistical	Full, OLS
6	STEPWISE (5)	Indirect	Statistical	Stepwise, OLS
7	$b_0 + b_1$ ASSETS$_{it}$+ ...	Direct	Statistical	Full, OLS
8	STEPWISE (7)	Direct	Statistical	Stepwise, OLS
9	WLS (3)	Indirect	Statistical	Full, WLS
10	WLS (4)	Indirect	Statistical	Stepwise, WLS
11	WLS (5)	Indirect	Statistical	Full, WLS
12	WLS (6)	Indirect	Statistical	Stepwise, WLS
13	WLS (7)	Direct	Statistical	Full, WLS
14	WLS (8)	Direct	Statistical	Stepwise, WLS

Notes: $E(IASPROF_{it})$ = Expected IAS profit for firm i in year t
FINPROF$_{it}$ = Reported Finnish profit for firm i in year t
OLS = Ordinary Least Squares regression
STEPWISE (k) = Stepwise regression of model k
WLS (k) = Weighted Least Squares (robust) regression of model (k)
$b_0, b_1, \ldots$ = Regression parameters

While Model 2 is a simplification of the general adjustment Equation (1), the following model is its statistical counterpart. Accordingly, for the third prediction model IAS profit was regressed on all income statement and balance sheet items that are associated with the general adjustment formula *and* that are usually available from Finnish financial statement information.

Model 3

$$E(IASPROF_{it}) = b_0 + b_1FINPROF_{it} + b_2AFFDIV_{it} + b_3 DELLOAN_{it}$$
$$+ b_4DELREC_{it} + b_5DIRMAN_{it} + b_6EQTAX_{it}$$
$$+ b_7EXGAIN_{it} + b_8FIXCOST_{it} + b_9INVCHG_{it} \quad (4)$$
$$+ b_{10}RENT_{it} + b_{11}RESCHG_{it} + b_{12}REVFUND_{it}$$
$$+ b_{13}SHARES_{it} + b_{14}TAXDEPR_{it}$$
$$+ b_{15}UNFPLIAB_{it} + b_{16}VALDEB_{it} + b_{17}VALCRE_{it}$$

where

$b_0, \ldots, b_{17}$ = Parameters to be estimated with the OLS (Ordinary Least Squares) method
FINPROF$_{it}$ = Finnish net income [1]
AFFDIV$_{it}$ = Dividends from affiliated companies [8]

DELLOAN$_{it}$ = Delivery credit liabilities [3]
DELREC$_{it}$ = Delivery credit receivables [3]
DIRMAN$_{it}$ = Direct manufacturing expenses [14]
EQTAX$_{it}$ = Taxes deducted from shareholders' equity [4]
EXGAIN$_{it}$ = Exchange gains and losses [7]
FIXCOST$_{it}$ = Fixed expenses [14]
INVCHG$_{it}$ = Change in inventories [14]
RENT$_{it}$ = Rent and lease expenses [9, 11, 13]
RESCHG$_{it}$ = Net change in untaxed reserves [2]
REVFUND$_{it}$ = Revaluation fund [10]
SHARES$_{it}$ = Shares and other securities [6, 8]
TAXDEPR$_{it}$ = Depreciation of fixed assets [15]
UNFPLIAB$_{it}$ = Unfunded pension liabilities [12]
VALCRE$_{it}$ = Valuation items: credit [7, 10]
VALDEB$_{it}$ = Valuation items: debit[7]

(Numbers in brackets above indicate the component number of the adjustment Equation (1) to which each variable was assumed to be associated.[9])

An obvious problem of Model 3 is the large number of independent variables. However, because there were no sufficient grounds to exclude any of these variables a priori, we let the data decide which of them are significant. Therefore, the stepwise regression method was used to generate **Model 4,** which is a compressed version of Model 3.

Because Model 3 includes FINPROF$_{it}$ as an independent variable, it can be regarded as an enlarged statistical counterpart of Model 1. Respectively, Model 5 can be regarded as an enlarged statistical counterpart of Model 2. In addition to the IASCALC$_{it}$ variable, it includes those independent variables in Model 3, which were *not* used in the computation of IASCALC$_{it}$.

Model 5

$$E(\text{IASPROF}_{it}) = b_0 + b_1\text{IASCALC}_{it} + b_2\text{AFFDIV}_{it} + b_3\text{DELLOAN}_{it}$$
$$+ b_4\text{DELREC}_{it} + b_5\text{IASCALC}_{it} + b_6\text{RENT}_{it}$$
$$+ b_7\text{REVFUND}_{it} + b_8\text{SHARES}_{it} + b_9\text{UNFPLIAB}_{it}$$
$$+ b_{10}\text{VALCRE}_{it} + b_{11}\text{VALDEB}_{it} \qquad (5)$$

Analogously to Model 4, **Model 6** was considered as a stepwise compressed version of Model 5.

In addition to the OLS method, which was first applied for the estimation of the parameters for Models 3 through 6, a WLS (Weighted Least Squares) method was also considered.[10] The primary reason for the use of this method was the authors' experience from prior research projects which had shown that WLS estimation can produce significant improvement in the estimation results especially in the presence of outliers in the data. Accordingly, **Models 9, 10, 11, and 12** were considered as the WLS estimated counterparts of Models 3 through 6, respectively.

Prediction Models Based on the Direct Method

As was noted above, the direct method of IAS profit determination implies the recalculation of the firm's net income according to the IAS accounting rules. Unfortunately, because the detailed bookkeeping records needed for the recalculation are neither at the disposal of security analysts nor researchers, they are not able to apply the direct method as such. The best one can do in this situation is to resort to the summaries of those records, as measured by the aggregate income statement and balance sheet items such as total assets, current assets and liabilities, net sales, operating income, and so on. Accordingly, the following model was tested as a statistical counterpart to the direct method of IAS profit determination.

Model 7

$$
\begin{aligned}
E(IASPROF_{it}) = b_0 &+ b_1 ASSETS_{it} + b_2 EQUITY_{it} + b_3 FINAS_{it} \\
&+ b_4 FINPROF_{it} + b_5 FIXAS_{it} + b_6 INVENT_{it} \\
&+ b_7 LDEBT_{it} + b_8 OPMARG_{it} + b_9 OPPROF_{it} \\
&+ b_{10} PROFBRT_{it} + b_{11} RESERV_{it} + b_{12} SALES_{it} \\
&+ b_{13} SDEBT_{it}
\end{aligned} \tag{6}
$$

where

$$
\begin{aligned}
ASSETS_{it} &= \text{Total assets} \\
EQUITY_{it} &= \text{Owners' equity} \\
FINAS_{it} &= \text{Financial assets} \\
FINPROF_{it} &= \text{Net income} \\
FIXAS_{it} &= \text{Fixed assets} \\
INVENT_{it} &= \text{Inventories} \\
LDEBT_{it} &= \text{Long-term debt} \\
OPMARG_{it} &= \text{Operating margin (before depreciation)} \\
OPPROF_{it} &= \text{Operating profit (after depreciation)} \\
PROFBRT_{it} &= \text{Profit before taxes and net change in reserves} \\
RESERV_{it} &= \text{Untaxed reserves} \\
SALES_{it} &= \text{Net sales} \\
SDEBT_{it} &= \text{Short-term debt}
\end{aligned}
$$

Analogously to Models 4 and 6, we also considered **Model 8** as a stepwise compressed version of Model 7. Finally, **Models 13 and 14** served as the WLS estimated counterparts of Models 7 and 8, respectively.

DATA AND METHODOLOGY FOR EMPIRICAL TESTS

Data

The practice of voluntarily disclosing earnings based on International Accounting Standards began in the early 1980s in Finland, when some companies decided to serve international markets by providing them with earnings information which would be more easily interpreted than the financial statements based on Finnish accounting rules. A survey of the annual reports of all major Finnish firms from the 1980s revealed that at least 18 companies had disclosed IAS-based accounting information during the last ten years. The total number of the IAS earnings announcements by those firms amounts to 93, and all of them were consolidated earnings of the group in question. The data used in this study comprised all these IAS earnings announcements.

In order to estimate the models and to test their predictive performance, the firms were split into two separate samples based on the following.

1. All firms were sorted in ascending order according to their size as measured by the amount of net sales in 1988.[11]
2. Starting from the top of the list, each firm was matched with another firm of approximately the same size and belonging to the same or closely related industry.
3. The firm in each pair for which more observations on the IAS earnings announcements were available was assigned to the estimation sample, and the other to the prediction sample.

As can be seen from Table 2, this matching produced estimation and prediction samples that have the following (desirable) properties: (a) they include firms from different industries; (b) while firms of different size are included in both samples, the average size of the firms in them is approximately equal; and (c) while more observations could be used for model estimation, the number of observations available for the prediction tests was still adequate.[12]

Empirical Tests

Theoretically, the selection of an optimal prediction model should be based on the decision maker's loss function.[13] Unfortunately, the loss function is

Table 2.   Estimation and Prediction Samples

Firm	Industry	Consolidated Net Sales 1988 (Million FIM)	Number of Observations
Estimation Sample			
Farmos	Medical	830	5
Lohja	Construction Material	2,729	5
Amer	Diversified	2,740	8
Wärtsilä	Shipbuilding	5,961	8
Yhtyneet	Wood-processing	6,300	4
Valmet	Machinery	8,517	5
Kemira	Chemistry	9,227	10
Rauma	Wood-processing	9,720	7
Nokia	Diversified	21,819	6
Total		67,843	58
Prediction Sample			
Tietotehdas	Service	472	4
Partek	Construction material	4,766	3
Huhtamäki	Diversified	4,439	5
Rosenlew	Wood-processing	1,613	4
Serlachius	Wood-processing	7,596	3
Tampella	Machinery	4,987	3
S. Sokeri	Food	4,051	6
Enso	Wood-processing	9,798	4
Kesko	Trade	26,311	3
Total		64,033	35

unknown in empirical studies and, therefore, the prediction performance of different model candidates is usually measured with prediction error measures which are assumed to be valid surrogates of the decision maker's loss function. In this study, the prediction performance of the models was analyzed with the following measures which differ significantly from each other with respect to their underlying assumptions on the theoretical loss function.[14]

1. Mean Error:

$$\text{ME} = (1/N) \sum_i \sum_t [E(\text{IASPROF}_{it}) - \text{IASPROF}_{it}] \qquad (7)$$

2. Mean Absolute Error:

$$\text{MAE} = (1/N) \sum_i \sum_t |[E(\text{IASPROF}_{it}) - \text{IASPROF}_{it}]| \qquad (8)$$

3. Mean Percentage Error:

$$\text{MPE} = (1/N) \sum_i \sum_t 100 * \frac{[E(\text{IASPROF}_{it}) - \text{IASPROF}_{it}]}{|\text{IASPROF}_{it}|} \qquad (9)$$

4. Mean Absolute Percentage Error:

$$\text{MAPE} = (1/N) \sum_i \sum_t 100 * \frac{|[E(\text{IASPROF}_{it}) - \text{IASPROF}_{it}]|}{|\text{IASPROF}_{it}|} \tag{10}$$

5. Root Mean Square Error:

$$\text{RMSE} = \{(1/N) \sum_i \sum_t [E(\text{IASPROF}_{it}) - \text{IASPROF}_{it}]^2\}^{1/2} \tag{11}$$

where

$E(\text{IASPROF}_{it}) = $ Predicted IAS profit for firm i in year t
$\text{IASPROF}_{it} = $ Actual IAS profit for firm i in year t
$N = $ Number of observations (= number of firms * number of years)

All error metrics listed above have at least one common problem—their sensitivity to individual outliers in the distribution of prediction errors. The problem is especially notorious for the relative error measures (MPE and MAPE) if the actual IAS profit for any firm in any year happens to be near zero. To avoid this problem, the central tendencies of the prediction errors were also defined with the medians.[15] Consequently, the following statistics were computed as additional error measures for this study.

6. Median Error: MDE
7. Median Absolute Error: MDAE
8. Median Percentage Error: MDPE
9. Median Absolute Percentage Error: MDAPE

Finally, the statistical test which were used to examine the significance of the differences between the prediction abilities of the models were the following.

1. The nonparametric *Friedman's Two-Way Analysis of Variance* (Siegel 1956, 166-73) was used to test overall differences between the prediction abilities of the model candidates.
2. The parametric *Paired T-Test* (Wonnacott and Wonnacott 1990, 268-72) was used for pairwise comparisons of the models with measures based on the means.
3. The nonparametric *Paired Wilcoxon's Test* (Siegel 1956, 75-83) was used for pairwise comparisons of the models with measures based on the medians.

EMPIRICAL RESULTS

Model Performance in the Estimation Sample

A summary of the performance of the prediction models in the estimation sample appears in Table 3. Although no parameters were to be estimated for Models 1 and 2, the relevant statistics were computed also for these two models in order to show their performance in this sample.[16]

The RMSE and R^2 statistics in the table reveal extremely poor performance for the naive Model 1 (FINPROF$_{it}$). The negative R^2 indicates that the sum of squared residuals for this model was even larger than the total sum of squares. This implies that the mean IAS profit gave, on average, better predictions for the sample firms than their Finnish profit figure, which can be explained by the expected difference in the levels of the two earnings figures.

The performance of Model 2 (IASCALC$_{it}$) is moderate in the estimation sample. The R^2 indicates that by adjusting the Finnish earnings number for the four differences between the Finnish and IAS accounting rules, the coefficient of determination rises from -5% to 76%. This suggests that these adjustments are very important, indeed.

At the other extreme one can see that Models 11 and 13 have the smallest RMSE and highest R^2. An explanation for their extremely high coefficient of determination (99%) can be found by looking at the number of observations: for example, in the estimation of Model 13 the WLS algorithm deleted 11 observations (19%) from the original sample as "outliers," and the final model was estimated from only 47 observations which the algorithm found to fit the model best. Similar tendencies to delete outliers can be seen also for Models 11 and 12 which explains their (very) good relative performance in the estimation sample.

It is also noteworthy that among the OLS estimated models, Models 3 and 4 show superior performance. Their coefficients of determination are about 90%, and due to the differences in the degrees of freedom, the statistics are only marginally better for Model 3 which has 17 regressors than for Model 4 which has 6.

The differences in the R^2 between the full and stepwise compressed models on one hand, and between the OLS and WLS estimated models on the other hand, were further analyzed by performing formal pairwise tests for their residuals (see Green 1978, 46-8). The results from these tests are reported in Table 4.

In brief, the findings of these tests suggest that, because difference in the sum of squared residuals of the full models (3, 5, and 7) and their stepwise-compressed counterparts (4, 6, and 8, respectively) was insignificant across all three comparisons, it might not be useful to retain the additional regressors of the full models. Moreover, the residuals of the WLS models (9-14) turned out to be significantly smaller than those of their OLS counterparts (3-8).

Table 3.   Summary of Estimation Results

Model	NIV	NOB	RMSE	Adj. R^2	Prob(F)
1	1	58	290533	−.0528	>.100
2	1	57	138814	.7635	<.001***
3	17	58	89365	.9004	<.001***
4	6	58	89385	.9003	<.001***
5	11	57	100437	.8762	<.001***
6	4	57	98861	.8801	<.001***
7	13	58	132863	.7798	<.001***
8	4	58	128616	.7937	<.001***
9	17	55	25149	.9879	<.001***
10	6	57	40126	.9667	<.001***
11	11	48	21339	.9894	<.001***
12	4	50	27407	.9827	<.001***
13	13	47	22926	.9919	<.001***
14	4	57	55250	.9094	<.001***

Notes: NIV = Number of independent variables; NOB = Number of observations used in the estimation; RMSE = Root Mean Square Error, adjusted for degrees of freedom; Adj. R^2 = Coefficient of determination, adjusted for degrees of freedom; Prob(F) = Marginal significance level of the F-ratio; *** Significant at 0.1% level.

Table 4.   Pairwise Comparison of Full versus Compressed Models and OLS versus WLS Models in the Estimation Sample

Models Compared	Type	SSE	Df	F-ratio	Prob(F)
3	FULL	3.1945E+11	40	1.002	>.100
4	COMP	4.0747E+11	51		
5	FULL	4.5394E+11	45	0.769	>.100
6	COMP	5.0822E+11	52		
7	FULL	7,7672E+11	44	0.630	>.100
8	COMP	8.7673E+11	53		
9	WLS	2.3402E+10	37	156.022	<.001***
3	OLS	3.1944E+11	40		
10	WLS	8.0505E+10	50	203.074	<.001***
4	OLS	4.0747E+11	51		
11	WLS	1.6393E+10	36	106.766	<.001***
5	OLS	4.5394E+11	45		
12	WLS	3.3801E+10	45	90.229	<.001***
6	OLS	5.0822E+11	52		
13	WLS	1.7346E+10	33	131.337	<.001***
7	OLS	7.7672E+11	44		
14	WLS	1.5873E+11	52	235.211	<.001***
8	OLS	8.7673E+11	53		

Notes: FULL = Full regression model; COMP = Compressed model through stepwise regression; WLS = Model estimated with the Weighted Least Squares method; OLS = Model estimated with the Ordinary Least Squares method; SSE = Sum of Squared Errors; Df = Degrees of freedom; F-ratio = (SSE$_2$ − SSE$_1$)/SSE$_1$ * Df$_1$/(Df$_2$ − Df$_1$); Prob(F) = Marginal significance level of the F-statistic; *** Significant at 0.1% level.

Model Performance in the Prediction Sample

The performance of the models in terms of the mean and median prediction errors in the *hold-out prediction sample* is shown in Table 5. The table reveals that the models with the best prediction performance according to different measures are the following.

ME:	Model 14 (10061)
MAE:	Model 2 (84241)
MPE:	Model 6 (-2.6%)
MAPE:	Model 2 (48.7%)
RMSE:	Model 2 (132364)
MDE:	Model 3 (2654)
MDAE:	Model 2 (37181)
MDPE:	Model 3 (0.6%)
MDAPE:	Model 2 (25.6%)

Model 2 was thus ranked first by five error measures, Model 3 by two error measures, and Models 6 and 14 by one measure.

Results from pairwise tests of the best performing models against each of their rivals are reported in Table 6 separately for each of the nine error measures.

The findings from Table 6 can be summarized as follows.

Error Measure	Best Model	Number of Rivals Significantly Outperformed
ME	Model 14	7
MAE	Model 2	8
MPE	Model 6	6
MAPE	Model 2	9
RMSE	Model 2	6
MDE	Model 3	4
MDAE	Model 2	8
MDPE	Model 3	4
MDAPE	Model 2	8

It turns out that, when Model 2 was ranked first by an error measure, the average number of rivals which it was able to significantly outperform was 7.8. The corresponding numbers are 7, 6, and 4 for Models 14, 6, and 3, respectively.

While the above results focused on pairwise tests of the best performing models according to individual error measures, the Friedman's Two-Way Analysis of Variance tests reported in Table 7 show the consistency of overall

Table 5.   Mean and Median Prediction Errors

Model	Mean Prediction Errors				
	ME	*MAE*	*MPE*	*MAPE*	*RMSE*
1	−162758	177151	−42.6	59.5	256462
2	−29011	84241	9.8	48.7	132364
3	−28253	144053	8.2	67.5	259142
4	−15619	131259	49.4	84.6	226387
5	−76153	133588	−84.9	124.7	198040
6	−35664	107640	−2.6	54.1	164511
7	−27660	107192	−35.1	102.4	147461
8	25239	87663	13.4	86.0	135140
9	−60421	148297	−53.5	106.3	223786
10	−29137	123469	33.8	71.0	209664
11	−49067	131748	−33.4	84.7	201726
12	−46620	114948	−17.6	65.3	173165
13	−23350	149198	−11.7	90.9	203769
14	10061	90848	7.5	82.8	138688

Model	Median Prediction Errors			
	MDE	*MDAE*	*MDPE*	*MDAPE*
1	−72650	91270	−53.7	60.4
2	−13873	37181	−8.1	25.6
3	2654	68214	0.6	52.9
4	17706	51712	14.9	48.9
5	−51330	65728	−43.3	55.9
6	−13127	40346	−13.6	33.5
7	−36027	68578	−22.9	40.3
8	12387	53457	11.1	35.3
9	−27721	84552	−21.5	53.4
10	10862	44909	9.6	34.3
11	−18499	53690	−20.1	50.5
12	−24825	49126	−18.5	41.5
13	−19807	95492	−29.4	56.4
14	7039	52229	5.0	31.6

Note: ME = Mean Error; MAE = Mean Absolute Error; MPE = Mean Percentage Error; MAPE = Mean Absolute Percentage Error; RMSE = Root Mean Square Error; MDE = Median Error; MDAE = Median Absolute Error; MDPE = Median Percentage Error; MDAPE = Median Absolute Percentage Error.

rankings when different prediction models and error measures are considered simultaneously.

The test statistics on the bottom of Table 7 indicate that the null hypothesis (that the rank order numbers for the prediction models came from the same population) can be rejected at levels of under 1% both for the mean-based and for the median-based error measures. The Kendall's coefficients of concordance on the far right (.499 and .777) show that model rankings by the median-based

Table 6.   Marginal Significance Levels for Pairwise Comparisons of Best
Predicting Models with Their Rivals

	Marginal Significance Levels for Mean Prediction Errors (Paired T-Tests)				
Model	*ME*	*MAE*	*MPE*	*MAPE*	*RMSE*
1	<.001***	.002**	.013*	>.100	.011*
2	.011*	—	>.100	—	—
3	>.100	>.100	>.100	.089	>.100
4	>.100	>.100	.015*	.033*	>.100
5	.002**	.007**	.009**	.013*	.033*
6	.041*	.044*	—	>.100	.028*
7	<.001***	.037*	>.100	.009**	>.100
8	<.001***	>.100	>.100	.024*	>.100
9	.073	.031*	.060	.024*	>.100
10	>.100	>.100	.031*	>.100	>.100
11	.062	.011*	.006**	.003**	.025*
12	.016*	.016*	.016*	.002**	.022*
13	>.100	<.001***	>.100	.003**	<.001***
14	—	>.100	>.100	.032*	>.100

	Marginal Significance Levels for Mean Prediction Errors (Paired Wilcoxon's Tests)			
Model	*MDE*	*MDAE*	*MDPE*	*MDAPE*
1	.005**	<.001**	.004**	.008**
2	>.100	—	>.100	—
3	—	>.100	—	>.100
4	>.100	.089	.075	.071
5	.041*	.006**	.003**	.003**
6	>.100	.041*	>.100	.074
7	>.100	.006**	>.100	<.001***
8	.039*	>.100	>.100	.028*
9	.046*	.012*	.018*	.007**
10	>.100	>.100	>.100	>.100
11	>.100	.043*	.041*	.006**
12	>.100	.012*	>.100	.002**
13	>.100	<.001***	>.100	<.001***
14	>.100	>.100	>.100	.059

Notes: * Significant at 5% level.
         ** Significant at 1% level.
         *** Significant at 0.1% level.

prediction errors were more consistent across different error measures than
the rankings given by the mean-based measures. Nevertheless, because both
tests indicate significant consistency across the error measures, the rank order
sums (RANKSUM) can be used for overall ranking of the models. They show
the superiority of Model 2 for the mean-based errors (see the upper panel)
as well as for the median-based errors (lower panel), and the findings thus fall
in line with the pairwise tests discussed above.[17]

Table 7. Model Ranks Based on Mean and Median Prediction Errors

Model	Ranks Based on Mean Prediction Errors					
	RANKSUM	*ME*	*MAE*	*MPE*	*MAPE*	*RMSE*
2	14	7	1	4	1	1
14	16	1	3	2	7	3
6	22	9	5	1	2	5
8	24	4	2	6	10	2
12	33	10	6	7	4	6
7	35	5	4	0	12	4
3	39	6	11	3	5	14
10	40	8	7	9	6	10
13	41	3	13	5	11	9
4	42	2	8	12	8	12
11	45	11	9	8	9	8
1	55	14	14	11	3	13
5	58	13	10	14	14	7
9	61	12	12	13	13	11

Model	Ranks Based on Median Prediction Errors				
	RANKSUM	*MDE*	*MDAE*	*MDPE*	*MDAPE*
2	11	6	1	3	1
14	12	2	6	2	2
10	14	3	3	4	4
6	16	5	2	6	3
8	21	4	7	5	5
3	22	1	10	1	10
4	27	7	5	7	8
12	29	10	4	8	7
11	34	8	8	9	9
7	40	12	11	11	6
9	44	11	12	10	11
5	47	13	9	13	12
13	48	9	14	12	13
1	55	14	13	14	14

Notes: Friedman's Test:

	χ^2	Df	Prob(χ^2)	W
Mean Prediction Errors	32.5	13	.002	.499
Median Prediction Errors	50.5	13	<.001	.777

Prob (χ^2) = Marginal significance level of χ^2 statistic; W = Kendall's coefficient of concordance.

The tests concerning the predictive performance of full versus stepwise compressed and OLS versus WLS estimated models are shown in Table 8. Because the corresponding tests in the estimation sample (Table 4) were based on the Sum of Squared Errors, the tests shown in Table 8 were based on analogous prediction errors (Mean Square Error and Median Square Error).

Table 8. Pairwise Comparison of Full versus Compressed Models and OLS versus WLS Models in the Prediction Sample

Models Compared	Type	Marginal Significance Levels	
		Mean Square Error (Paired T-Test)	Median Square Error (Wilcoxon's Test)
3	FULL		
4	COMP	>.100	>.100
5	FULL		
6	COMP	.072	.075
7	FULL		
8	COMP	>.100	>.100
9	WLS		
3	OLS	>.100	>.100
10	WLS		
4	OLS	>.100	>.100
11	WLS		
5	OLS	>.100	>.100
12	WLS		
6	OLS	.044*	.036*
13	WLS		
7	OLS	.011*	>.100
14	WLS		
8	OLS	>.100	>.100

Notes: FULL = Full regression model; COMP = Compressed model through stepwise regression; WLS = Model estimated with the Weighted Least Squares method; OLS = Model estimated with the Ordinary Least Squares method; * Significant at 1% level.

On the whole, the results for the full versus compressed model comparisons do not indicate significant differences in the predictive performance of the two model types. The results in Table 8 thus are consistent with those in Table 4 which suggested that the additional regressors in the full models do not produce significant improvement in model performance.

As regards the WLS versus OLS comparisons, the results in Table 8 show that OLS Models 6 and 7 tend to outperform their WLS counterparts (12 and 13) significantly. It is especially noteworthy that the significant superior performance of the WLS models in the estimation sample (Table 4) thus does not seem to be generalizable to the prediction sample.

SUMMARY AND CONCLUSIONS

The main purpose of this paper was to test alternative models for predicting IAS profits from the financial statements based on (non-IAS) Finnish accounting rules. Such model(s) should be useful to security analysts, accountants, as well as to researchers. Finland provides a particularly interesting case for this kind of study because of important differences between International Accounting Standards and Finnish accounting rules.

Altogether 14 prediction models based on the indirect and direct methods of IAS profit determination were examined in this study. Besides the methodological background, the models differed from each other with respect to their mathematical nature and statistical estimation technique. The results were as follows: Statistical regression models employing the Weighted Least Squares estimation showed, in general, superior performance in the estimation sample. It also turned out that the naive prediction of no difference between the IAS and Finnish profits performed very poorly in the estimation sample.

When the estimated models were tested in a hold-out prediction sample, a relatively simple analytical model based on adjusting the Finnish profit for four important differences between IAS and Finnish accounting rules, tended to outperform statistical regression models irrespective of their number of regressors or the method of parameter estimation. Interestingly, the results showed that the superiority of the WLS estimated models in the estimation sample did not, in general, carry to the prediction sample. Finally, it can be noted that the expectations on the poor performance of naive predictions were confirmed also in the prediction sample.

In conclusion, the findings of this study suggest that statistical regression models can do little in improving IAS profit predictions beyond what is obtained through adjusting the reported Finnish profit for a few important differences in the accounting rules. If more accurate predictions are required, they are likely to be found from further adjustments of the earnings for the remaining differences between IAS and Finnish accounting rules.

APPENDIX

Deriving A Shortcut Analytical Formula for Estimating the IAS Profit from Finnish Financial Statement Information

As was noted in the body of the text, the following components of the general Equation (1) were taken into account for the derivation of IASCALC: (1) net change in untaxed reserves, (2) taxes deducted from shareholders' equity, (3) an adjustment for manufacturing overhead, and (4) an adjustment for depreciation method. Because the first two of these are, in general, readily available from published financial statements, making adjustments for them should not pose any problems. However, the adjustments for manufacturing overhead and depreciation are not quite as straightforward, as shown below.

Adjustment for Depreciation

A dynamic model is needed to estimate the IAS straight-line depreciation from the geometrically declining Finnish depreciation. We have to make some assumptions concerning the investment schedule and estimated useful asset lives. The following (simplifying) assumptions were made for the analysis.

1. The investment have grown at a steady growth rate which is approximated by the nominal growth rate of the GNP.
2. The companies have taken maximum declining balance depreciation allowed by the tax rules (30% for machinery and 9% for buildings).
3. The useful asset lives for machinery and buildings are determined according to the U.S. ACRS system.

The Finnish depreciation is given by the following formula.

$$d_{FIN}(t) = p^*(A(t-1) + I(t)), \quad A(t) = (1-p) * (A(t-1) + I(t)) \quad \text{(A1)}$$

where

$d(t) =$ depreciation for period $(t, t-1)$,
$p =$ depreciation percentage,
$A(t) =$ depreciable assets at time t,
$I(t) =$ net investments during period $(t, t-1)$

At the end of the asset life, the remaing book-value of the asset will be expensed.
The IAS straight-line depreciation is given by the formula.

$$d_{IAS}(t) = A(O)/N \quad \text{(A2)}$$

where

$N =$ asset life,
$A(O) =$ original cost of the asset.

At the end of the asset life, the book-value of the asset, net of cumulative depreciation, will be automatically zero.

Investments were assumed to be growing according to the rule:

$$I(t) = (1 + g)^{t-1}I(1) \quad \text{(A3)}$$

where $g =$ constant growth rate (nominal GNP growth rate in this model).

Given the definitions of $d_{FIN}(t)$ and $d_{IAS}(t)$ and the assumption in (A3), we can calculate the ratio of the annual depreciation along a constant growth path.

Using the properties of geometric series and mathematical induction, it can be shown that at the end of the asset life we have

$$c = \frac{d_{IAS}(t)}{d_{FIN}(t)} = \frac{[(1+g)^N - 1]/(gN)}{p[(1+g)^{N-1}+(1+g)^{N-2}(1-p)+ \ldots +(1-p)^{N-1}]+(1-p)^N} \quad (A4)$$

With a constant growth rate, c is the steady state value of the ratio between IAS straight-line depreciation and Finnish geometrically declining depreciation. Given the Finnish depreciation, we get the (estimate of) IAS depreciation by simply multiplying the Finnish depreciation by the coefficient c.

As an estimate of g, we used 11.3% p.a. which was the nominal growth rate of the Finnish GNP during 1980-1988. Thirty-two years were used to approximate the useful life of buildings and 7 years to approximate that of machinery and equipment. The selections of the asset lives were based on the table of asset lives of different classes of assets for the Accelerated Cost Recovery System (ACRS) under the U.S. 1986 Tax Reform Act. With these parameter values, the coefficient values were $c_B = 0.602$ for buildings and $c_M = 0.905$ for machinery and equipment.

The adjustment formula (A40) was used only for machinery and equipment and for buildings which in most cases are the most important depreciated asset classes. Depreciations of other assets were taken as they were reported in the financial statements. We further assumed that the difference between the Finnish tax depreciation and the straight-line depreciation was attributable to the depreciation of machinery and equipment and buildings. The computation of total IAS depreciation for firm i in year t was then as follows:

$$\begin{aligned}
IASDEPR_{it} = c_B &* \{T152 + [T152/(T152 + T153)] * T158\} \\
&+ c_M * \{T153 + [1 - T152/(T152 + T153)] * T158\} \\
&+ (T151 + T154 + T155 + T156) \quad (A5)
\end{aligned}$$

where

c_M = Adjustment coefficent for machinery and equipment
c_B = Adjustement coefficent for buildings
T151 = Depreciation: land
T152 = Depreciation: buildings
T153 = Depreciation: machinery and equipment
T154 = Depreciation: intangible rights
T155 = Depreciation: other long-term expenditures
T156 = Depreciation: goodwill
T158 = The difference between the depreciation according to plan and the tax depreciation

Adjustment for Manufacturing Overhead

In order to adjust the inventory change reported in the Finnish income statement for manufacturing overhead, the following formula was applied:

$$\frac{INDMAN_{it}}{DIRMAN_{it}} * (T31/(T30 + T31 + T32)) * T141 \qquad (A6)$$

where

INDMAN = indirect manufacturing costs = T142 + ... + T146 +
$\quad c_B * (T152 + (T152/(T152 + T153)) * T158)$
$\quad c_M * (T153 + (1 - T152/(T152 + T153)) * T158)$
DIRMAN = direct manufacturing costs = T136 + ... + T139

in which

> T130 = Raw materials
> T131 = Finished products
> T132 = Other inventories
> T136 = Raw material (purchases)
> T137 = Variable labor expenses
> T138 = Social security expenses (variable)
> T139 = Other variable expenses
> T141 = Change in inventories
> T142 = Fixed labor expenses
> T143 = Social security expenses (fixed)
> T144 = Rent expense
> T145 = Lease expense
> T146 = Other short-term fixed expenses

T152 ... T158 = As defined above

One problem of the above equation is that other than manufacturing overhead can be included in the numerator of the first team. Another problem is encountered when the firm does not disclose variable and fixed expenses separately in which case it is very difficult (if not impossible) to make the appropriate adjustment for overhead.

ACKNOWLEDGMENT

This paper was written while the second author was visiting the EIASM (European Institute for Advanced Studies in Management), Brussels. The study is part of a larger research project that examines the extent and implications of income smoothing practices in Finnish firms. Financial support for the project was provided by the Foundation for Economic Education (Finland) which is gratefully acknowledged. The authors are also grateful for the helpful comments of an anonymous referee.

NOTES

1. The term "prediction" is used in this paper in the sense of deriving a contemporaneous estimate of IAS profit from available domestic financial statements. The term is thus *not* used in the sense of forecasting future IAS profit from current financial statement information.

2. This view is based on the authors' teaching experience at the Helsinki School of Economics where International Accounting Standards are included in the third-year undergraduate course on financial accounting. The experience has shown that the students (and the teachers as well) find it very difficult to derive exact IAS profit from Finnish financial statement information even for a simplified example.

3. For a review of the income smoothing literature, see Ronen and Sadan (1981). External motives of income smoothing typically relate to the enhancement of the share value of the firm through supporting a higher level of dividends, through reducing the market risk of the firm, and through minimizing income taxes.

4. While the average index measuring the conformity of national accounting rules with International Accounting Standards in the IASC (1988) survey was 82%, the index for Finland was only 12%.

5. See Kinnunen (1989), who showed that depreciation and changes in untaxed reserves reduced, on average, more than 90% of the variance of annual earnings change in a sample of 33 listed Finnish firms.

6. A more detailed comparison of IAS and Finnish accounting standards can be found in Blomquist (1987), Guarnieri (1987), Hällström (1987), Riistama (1987), and Viljanen (1987). See Salmi, Virkkunen and Helenius (1987) for an English presentation of the Finnish accounting rules, and Salmi (11978) for a discussion of their theoretical foundations.

7. The untaxed reserves allowed by Finnish regulations include inventory reserve, reserve for bad debts, reserve for warranty repairs, investment reserve, and operational reserve. In manufacturing firms, the most important of these is the inventory reserve, the maximum amount of which is 40% of the inventories valued at the lower of direct (FIFO) cost or market value.

8. It can be noted that these methods are analogous to the direct and indirect methods of cash flow computation (see, e.g., Mahoney, Sever, and Theis 1988).

9. Because none of the firms analyzed in this study disclosed income directly added to shareholder's equity, we had no variable measuring component #5 in Model 3.

10. The WLS method used in this study was based on Andrew's Sine-algorithm available in the statistical SOLO-package developed by BMDP Statistical Software, Inc.

11. Because observations of net sales in 1988 were not available from two firms, they were taken from preceding years.

12. It should be noted that before any empirical tests were performed, the data was visually checked for potential outliers. This was done by plotting the dependent IASPROF$_{it}$ variable against each independent variable, one at a time, in the regression Models (3, 5, and 7). Because three obvious outliers were revealed, one relating to the dependent variable and the other two to different independent variables (FINPROF$_{it}$ and AFFDIV$_{it}$), they were set to missing values for subsequent computer runs. Consequently, there will appear slightly different numbers of observations across the models, depending on which variables they employ.

13. The term "loss" refers to the difference between the return which could be earned if the value of the variable being predicted (IAS profit) were known with certainty *and* the return which

the decision maker actually can achieve when he uses uncertain predictions as a basis of his (her) decision. If the decision maker is rational, he (she) will then choose that prediction model which minimizes the sum of expected loss plus the direct costs incurred from using the model (see, e.g., Demski and Feltman (1972) for a closer theoretical discussion on forecast evaluation).

14. One should note that the divergence of assumptions of the error measures is desirable simply because a set of measures with different assumptions obviously has a greater descriptive validity with respect to the unknown theoretical loss function than a set of measures with similar assumptions.

15. Another way to avoid this problem would be to impose a truncation rule for individual errors, for example all errors outside the $\pm 500\%$ interval are set to $\pm 500\%$. This approach has, however, two problems: (1) the limits have to be set subjectively, and (2) the error distribution is no longer continuous.

16. Because we are in search for the best model in terms of prediction ability, the analysis of the sign and the statistical significance of individual parameter estimates are not as important as in explanatory studies trying to verify a priori hypotheses between the dependent and independent variables. Because of this, and in order to save space, the parameter estimates (and respective t-values) have been omitted from Table 3.

17. Because the overall prediction ability results in Table 7 showed the superiority of Model 2, its validity was examined with the following simple test. The prediction Equation (2) for Model 2 suggests that $b_0 = 0$ and $b_1 = 1$ in the following regression:

$$\text{IASPROF}_{it} + b_0 + b_1 \text{IASCALC}_{it} + e_{it} \qquad (12)$$

In order to test the hypotheses on b_0 and b_1, Equation (12) was estimated with the OLS method from the aggregate sample. In brief, the results gave no support for the rejection of the null hypotheses that $b_0 = 0$ $b_1 = 1$ in Equation (12), which was indicated by the low t-statistics and corresponding insignificant marginal significance levels.

REFERENCES

Blomquist, L. 1987. Kansainvälisen käytännön mukainen tilinpäätös [Financial statements according to the international accounting practice]. *Tilintarkastus* 6: 393-97.

Demski, J.S., and G.A. Feltman 1972. Forecase evaluation. *The Accounting Review* (July): 533-548.

Green, P.E. 1978. *Analyzing Multivariate Data*. City: Dryden Press.

Guarnieri, S. 1987. Liitetiedot kansainvälisen käytännön mukaan [Appendices of financial statements according to the international accounting practice]. *Tilintarkastus* 6: 404-7.

Hällström, af C. 1987. Vaihto-omaisuudesta kansainvälisen käytännön mukaan [On the inventories according to international accounting practices]. *Tilintarkastus* 6: 398-403.

International Accounting Standard Committee. 1988. *Survey of the Use and Application of International Accounting Standards*. London: IASC.

Kinnunen, J. 1989. The income smoothing effects of depreciations and untaxed reserves in listed Finnish firms. *The Finnish Journal of Business Economics* 4: 293-305.

Mahoney, J., M. Sever, and J. Theis. 1988. Cash flow: FASB opens the floodgates. *Journal of Accountancy* (May): 26-38.

Riistama, V. 1987. Tekeekö yhtenäistyvä kansainvälinen tilinpäätöskäytäntö oman ajattelun tarpeettomaksi? [Does the harmonization of international accounting practice make your own thinking useless?]. *Tilintarkastus* 6: 419-26.

Ronen, J., and S. Sadan. 1981. *Smoothing Income Numbers: Objectives, Means, and Implications.* Redding, MA: Addison-Wesley.

Salmi, T. 1978. A comparative review of the Finnish expenditures revenue accounting. EIASM working paper 78-2 Brussels.

Salmi, Virkkunen, & Helenius Ky [Coopers & Lybrand Oy]. 1987. *An Investor's Guide to Finnish Reporting Requirements and Taxation.* Helsinki.

Siegel, S. 1956. *Nonparametric Statistics for the Behavioral Sciences.* New York: McGraw-Hill.

Viljanen, J. 1987. IAS-tilinpäätöksen laadinnan pääperiaatteet [The principles of preparing financial statements based on the international accounting standards]. *Tilintarkastus* 6: 408-18.

Wonnacott, T.H., and R.J. Wonnacott. 1990. *Introductory Statistics for Business and Economics.* New York: Wiley.

RECENT INNOVATIONS IN GERMAN ACCOUNTING PRACTICE THROUGH THE INTEGRATION OF EC DIRECTIVES

Timothy S. Doupnik

ABSTRACT

With the integration of EC Directives into German accounting law in 1985, significant changes have been made in German financial reporting that bring it more in line with Anglo-American practice. However, important differences continue to exist. This paper outlines the recent innovations in German accounting law and compares the current system with accounting principles generally accepted in the United States. The primary purpose of this paper is to provide a relatively comprehensive picture of current German accounting practice which might be of use to accounting scholars in their teaching and research efforts. The reunification of Germany and the important role a united Germany undoubtedly will have in a post-1992 Europe make the study of German accounting all the more relevant.

Advances in International Accounting,
Volume 5, pages 75-103.
Copyright © 1992 by JAI Press Inc.
All rights of reproduction in any form reserved.
ISBN: 1-55938-415-8

INTRODUCTION

Historically, German accounting and financial reporting has been considerably different from the system used in the United States. Classification studies have consistently placed Germany in a class different from the United States (see, for example, Mueller 1968; AAA 1977; Nair and Frank 1980; Nobes and Parker 1981). In a study of 1979 and 1980 annual reports from 10 countries, Choi and Bavishi (1982) found that of 32 material accounting practices followed in the United States, only 18 were followed by companies in Germany.

With the integration of the European Community's (EC) Fourth, Seventh, and Eighth Directives into German law in 1985, German accounting practices have made considerable movement in the direction of Anglo-American practice. However, differences continue to exist. The objective of this paper is to describe the changes that have been made in German accounting practices and point out the differences between German and U.S. accounting practices which remain.

The primary motivation for this paper is to provide current information on German accounting practices that might be of use to international accounting scholars in their teaching and research efforts. This paper should also be of interest to practitioners and investors who might be interested in doing business in Germany. In 1988, West Germany was the third most popular location of United States foreign direct investment behind Canada and the United Kingdom. With the impending integration of markets as a result of Europe 1992 and the dominant role a unified Germany is expected to play in that common market, the study of German accounting practices becomes all the more timely.

The analysis begins with an examination of basic principles underlying corporate financial reporting in the two countries. Many of the differences in accounting practices that continue to exist can be traced to fundamental differences in the relative importance placed on basic principles. The components, formats, and classification of financial statements are then described and compared with those of the United States. This is then followed by a comparison of some of the more important valuation and measurement rules, including consolidation procedures.

The Accounting Directives Law of 1985

The Accounting Directives Law (*Bilanzrichtliniengesetz*) came into effect on January 1, 1986. With its publication, the difficult process of transforming the Fourth EC Directive (Accounting), Seventh EC Directive (Consolidated Accounts), and Eighth EC directive (Auditing) into German law was completed. Rather than writing separate laws to transform each of the EC directives into German law separately, the German Parliament decided to

Table 1. Major Innovations in German Accounting Directives Law of 1985

The following is a list of major changes in German accounting practice brought about by the integration of EC Directives into German law:

1. preparation of notes which become an integral part of the financial statements,
2. preparation of consolidated financial statements on a worldwide basis, that is, foreign subsidiaries can no longer be excluded,
3. use of the parent company or entity theory methods of consolidation (rather than the so-called German method),
4. requirement that all companies included in consolidated financial statements apply uniform valuation procedures, that is, GAAP conversion made if necessary,
5. elimination of unrealized intercompany losses on consolidation,
6. use of the equity method for investments in associated companies,
7. disclosure of comparative figures in the balance sheet and income statement,
8. disclosure of liabilities with a maturity of less than one year,
9. disclosure of the gross amount of fixed assets with accumulated depreciation shown separately,
10. obligatory accrual of deferred tax liabilities,
11. obligatory accrual of pension obligations,
    and
12. option to capitalize deferred tax assets.

integrate all three directives into one piece of legislation. The resulting Accounting Directives Law modifies 39 existing laws. Of most importance for accounting is the modification of the Commercial Code (*Handelsgesetzbuch—* HGB). With the passage of the Accounting Directives Law, the HGB has become the major authoritative pronouncement governing accounting practices in Germany, superceding the Stock Corporation Law (*Aktiengesetz—* AktG) of 1965.

The Accounting Directives Law is very comprehensive but not all inclusive. It is silent with regard to several very important accounting issues. These include methods of pension and lease accounting, translation of foreign currency financial statements, and cash flow information. With regard to pension accounting, German companies receive their direction from the tax law. Companies must refer to other sources including the academic and practitioner literature, however, in developing procedures for lease accounting and foreign currency translation. The significance of the innovations of the new law can be seen from the list of major changes shown in Table 1. Items 1-10 have been generally accepted accounting practice in the United States for more than 20 years. It should be noted that although most provisions of the Accounting Directives Law became effective in 1986, the preparation of worldwide consolidated financial statements did not become obligatory until 1990.

In addition to changes in valuation and disclosure requirements, consistent with the Fourth EC Directive, the Accounting Directives Law introduced the overriding principle of "true and fair view" into German accounting practice. The law states that the financial statements must, in compliance with accepted accounting principles, present a true and fair view of the net worth, financial position, and results of the company (HGB Sec. 264(2)).

This new orientation has found its way into the auditors' opinion. Previously the audit opinion merely stated that the financial statements comply with German legal requirements. In conjunction with this new principle, the audit opinion has been changed to state, in addition, that the financial statements present a "true and fair view" of the net worth, financial position, and results of the company.

Notwithstanding the changes brought about by the new law, differences continue to exist between German and United States accounting practices which can be partially traced to historical differences in orientation and application of general principles.

GENERAL PRINCIPLES OF ACCOUNTING

Both United States and German accounting practice are based on similar fundamental principles such as: historical cost, consistency, conservatism, and revenue realization. Although the basic principles are the same in the two countries, relative weights attached to the various principles differ which lead to potentially significant differences in accounting measurement and valuation.

In the United States the criteria of relevance and reliability lead to an overriding principle of "fair presentation." The auditors' opinion in the United States attests to the fact that the financial statements "present fairly in all material respects" the financial position, results, and cash flows of the company. Deviations from other general principles are made if necessary for fair presentation. For example, the percentage-of-completion method is used for long-term construction projects, in violation of the date-of-sale revenue realization principle, because it more fairly presents the results of operations for the current period.

Conservatism Principle

In Germany, the overriding principle historically has been the conservatism principle (*Vorsichtsprinzip*) and this situation continues under the new law. Options available within the law allow companies to record assets at values lower than would be acceptable in the United States and to record liabilities that would not be recorded in the United States. For example, German companies are allowed to use prime costing in determining the cost of inventories, that is, all overhead is expensed immediately (HGB Sec. 255(2)), and are allowed to accrue a liability for postponed repairs and maintenance expenditures that will be made in the following period (HGB Sec. 249(1)). It has been suggested that the general policy of German companies is to value assets as low as is acceptable under law and value liabilities as high as is acceptable, with the intent to report profit as low as possible (Jung, Isele, and Gross 1989, 93). This can be contrasted with the situation in the United States where companies generally are interested in reporting higher profit because of the perceived impact this has on stock prices.

The emphasis on the conversatism principle is at least partially attributable to the creditor protection (*Glaübigerschutz*) orientation in Germany, as opposed to the equity investor orientation in the United States Historically, debt has been more important than equity as a source of corporate funds in Germany. In an economy of $1.3 trillion, there are only 492 publicly traded companies in Germany (*Forbes* 1990, 123). The emphasis on creditor protection is also manifested in a more restrictive definition of assets than is found in the United States. Assets are defined as resources that can be severed from the company and sold.

The objective of reporting lower profit is further emphasized in Germany because of the *Massgeblichkeitsprinzip* which provides that the financial statements form the basis for taxation. As Lück (1983, 751) explains, if German tax law allows choice among different valuation methods which are also acceptable for financial reporting, then the method selected for financial reporting must also be used for tax purposes. A different method may be used for tax purposes only if tax law specifically requires that a method unacceptable for financial reporting be used. The practical effect of the *Massgeblichkeitsprinzip* is that, to reduce taxes, companies have an incentive to report income as low as possible in the corporate financial statements. This principle, which was developed in the nineteenth century to simplify the calculation of taxable income, appears to be uncontestable in Germany, and forced the legislature to include provisions in the Accounting Directives Law which directly conflict with the principle of true and fair view. For example, unplanned depreciation taken on fixed assets (due, for example, to technical obsolescence) or extraordinary write-downs of current assets generally must be reversed if the reason for them no longer exists (HGB Sec. 280(1)). However, such reversals need not be made if the lower valuation can be retained for tax purposes, and it is a prerequisite of such retention that the value be retained in the financial statements as well (HGB Sec. 280(2)).

Over the years, as a result of the German legislature attempting to achieve nonfiscal objectives through tax incentives, a "reverse" *Massgeblichkeitsprinzip* has developed (Lück 1983, 752). (A similar tax conformity rule exists in the United States only with regard to the use of LIFO.) German tax law allows companies to take extraordinary or "special" depreciation (*Sonderabschreibungen*) on investments in certain types of assets but only if the extraordinary depreciation is also reflected in the financial statements. This tax conformity rule can result in assets being valued below their net realizable value which violates the principle of true and fair view. The new law attempts to comply with the true and fair view doctrine and at the same time accommodate the tax conformity rule by creating a balance sheet heading that can be literally translated as "special account with a reserve component" (*Sonderposten mit Rücklageanteil*). This account is similar to a valuation adjustment account that might be used to reconcile LIFO inventory to current cost. It is discussed further in the section titled "Measurement of Liabilities."

The emphasis on conservatism in Germany allows companies to value assets at relatively low amounts and estimate liabilities at relatively high amounts creating so-called "hidden reserves" (*stille Reserven*). To a great extent companies view hidden reserves as an income smoothing tool; build them up in good years and dissolve them in bad years. For example, Heinen (1980, 279) states that the primary reason for creating hidden reserves is for dividend stabilization purposes. He goes on to mention that companies can fool their stockholders into believing that a poor profit year was actually good through the dissolution of hidden reserves, and that the use of hidden reserves for smoothing purposes can be easily misused by an irresponsible management (1980, 279). Another possible reason for a financial reporting policy of income minimization is to mitigate labor unions' demands for higher wages that might result from higher reported income. Labor unions have a greater impact on companies in Germany than in the United States as evidenced by the law requiring employee representation on the Supervisory Board (*Aufsichtsrat*).

The potential size that these hidden reserves can attain was dramatically reflected in Daimler Benz's 1989 annual profit which was expected to be somewhat lower than the DM1.7 billion earned in 1988 (*Der Spiegel* 1990). When 1989 profit was announced to be almost DM7 billion, a four-fold increase over 1988, Daimler's chairman Edzard Reuter was hailed as a magician capable of pulling rabbits out of the hat (1990, 114). As footnote 22 in the 1989 Annual Report indicates, this dramatic increase in profit was achieved by reversing (1) previously accrued pension liabilities, (2) write-downs of assets for potential losses, and (3) provisions for future plant alterations and maintenance. The reason given for these changes in accounting principles is as follows:

> Concomitant with the restructuring into an integrated high-technology group which operates all over the world, new accounting policies were applied to the consolidated financial statements for 1989. As a company engaged only in the automobile business, we traditionally used to pursue an accounting and valuation policy which scarcely, if at all, allowed the newly created high-technology group to be accurately assessed on an international scale. However, proper assessment of the earning power and financial strength of a company is of great significance when it comes to capital procurement. The intention of trading the Daimler-Benz share at the most important foreign stock exchanges requires a balance sheet structure and disclosure of stockholders' equity which meet internationally customary standards (Daimler-Benz 1989 Annual Report, 17).

Thus, at least one major company in Germany has changed its policy with regard to hidden reserves for the purpose of competing in international capital markets.

Hidden reserves and income smoothing represents one of the greatest differences in orientation between United States and German accounting practice. Numerous studies have attempted to show that smoothing occurs in the United States, but the results are inconclusive (Imhoff 1981). To the extent that smoothing does exist, it is not widely discussed in public. In contrast, in Germany, hidden reserves are openly discussed and textbooks even enumerate the ways in which they can be created and subsequently dissolved (see, for example, Heinen 1980, 279-285). The process of using options within the law to generate the desired amount of reported profit is referred to as *Bilanzpolitik* (literal translation: financial statement policy).

Other Principles

German accounting practice allows few, if any, deviations from the general principles of historical cost, conservatism, and revenue realization, whereas deviations are required in the United States in specific situations. For example, in the United States under SFAS 52, unrealized transaction-based foreign exchange gains are reported in current income even though this violates the realization and conservatism principles. Because of these violations, the accrual of unrealized foreign exchange gains is forbidden in Germany. In addition, the percentage-of-completion method, which deviates from the conservatism and date-of-sale revenue recognition principles, is not allowed in Germany except in the situation where a project is completed in contractually defined segments.

There are two "general valuation principles" enumerated in the new law, unknown in the United States, which lead to specific differences in United States and German accounting practice. The first of these is the "balance sheet continuity principle" (*Grundsatz der Bilanzidentität*) which requires that the values included in the opening balance sheet of one financial year must agree with those of the closing balance sheet of the previous year (HGB Sec. 252(1)). This precludes restatement of comparative balance sheets for changes in accounting or consolidation policy, such as was necessary in the United States as companies consolidated their previously unconsolidated finance subsidiaries under SFAS 94 in 1988. This would also preclude treating corrections of errors as adjustments to the beginning balance of retained earnings as required by APB Opinion 20.

The second item relates to the "principle of individual valuation" (*Grundsatz der Einzelbewertung*) which requires that assets and liabilities be valued at the balance sheet date on an item-by-item basis (HGB Sec. 252(1)). This precludes German companies from applying the lower-of-cost-or-market rule to marketable securities on a portfolio basis as is required in the United States. The potential result is that German companies report marketable securities at a lower value than is allowed in the United States, another source of hidden reserves.

The overriding constraint of materiality embedded in United States accounting practice is not found in Germany. Under previous law, strict observance of the letter of the law was required, that is, nondisclosure of required items could not be justified on the grounds of immateriality. The Accounting Directives Law has introduced the materiality concept with regard to certain disclosure requirements, but not as an overriding constraint. For example, Section 268 of the Law specifically indicates that disclosure of accrued post-balance sheet assets and liabilities must be made only if material.

COMPONENTS OF FINANCIAL STATEMENTS

Financial statements in Germany consist of a balance sheet (*Bilanz*), income statement (*Gewinn- und Verlustrechnung*), and the newly required notes to the financial statements (*Anhang*). The financial statements must also include a schedule of changes in long-term assets (*Anlagespiegel*), and are accompanied by a required management report (*Lagebericht*) and auditors' opinion (*Bestätigungsvermerk*). German companies are required to present both parent company only (*Jahresabschluss der A G*) and consolidated financial statements (*Konzernabschluss*) in their annual reports.

The new law is silent with regard to a statement of cash flows; it is neither required nor forbidden. Major German multinational companies tend to include cash flow information (*Mittelherkunfts- und Mittelverwendungsrechnung*) either in the notes or in the management report. Because it is not regulated by law, the form of this statement varies across companies and tends to differ from the form required under SFAS 95.

BALANCE SHEET CLASSIFICATION

The content and format of the balance sheet in Germany is regulated by law (see Table 2). This greatly increases the comparability across companies. Subdivisions of headings are permissible, new headings may be added if their contents are not covered by any prescribed items, and headings may be combined if they include immaterial amounts or clarity of presentation is thereby improved.

Assets

German accounting practice dichotomizes assets into *Umlaufvermögen* and *Anlagevermögen*. German companies preparing English language translations

Table 2. Balance Sheet Classification

Assets	Equities
A. SUSPENSE ACCOUNT	A. OWNERS' EQUITY
B. LONG-TERM ASSETS	I. Capital Stock
I. Intangible Assets	II. Additional paid-in capital
1. patents, trademarks, other legal rights	III. Reserves
2. goodwill	1. legal reserves
3. payments in advance	2. for treasury stock
II. Fixed Assets	3. required by corporate charter
1. land and buildings	4. other reserves
2. machinery and equipment	IV. Profit Carried Forward
3. other fixed assets	V. Current Net Income
4. payments in advance	B. SPECIAL ACCOUNT WITH A RESERVE COMPONENT
III. Financial Assets	C. ESTIMATED LIABILITIES
1. investments in subsidiaries	1. accrued pension and similar obligations
2. loans to subsidiaries	2. accrued tax liability
3. investments in associated entities	3. deferred tax liability
4. loans to associated entities	4. other accrued liabilities
5. long-term marketable securities	D. LEGAL OBLIGATIONS
6. other loans	1. bonds payable, of which convertible
C. CURRENT ASSETS	2. liabilities to financial institutions
I. Inventories	3. advance deposits
1. raw materials, supplies	4. accounts payable
2. work-in-progress	5. notes payable
3. finished goods	6. liabilities to subsidiaries
4. payments in advance	7. liabilities to associated entities
II. Receivables and Other Assets	8. other liabilities, of which taxes, of which in the area of social security
1. trade receivables entities	E. DEFERRED CREDITS
2. receivables from subsidiaries	
3. receivables from associated entities	
4. other assets	
III. Marketable Securities	
1. investments in subsidiaries	
2. treasury stock	
3. other marketable securities	
IV. Cash on hand, checks, bank accounts	
D. DEFERRED CHARGES	

Source: Handelsgesetzbuch Section 266.

of their annual reports generally translate the former as current assets and the latter as long-term or noncurrent assets. However, the German concepts are somewhat different from their English counterparts, being based on both time and function.

Umlaufvermögen (current assets) are defined as those assets not intended to serve the business on a continuous basis (Lück 1983, 1107). The classification is similar to the United States classification of current assets, comprised of inventories, receivables, marketable securities, and cash, but the two concepts are not identical in that German current assets include long-term receivables and exclude prepaid expenses. Prepaid expenses, both current and long-term, are included in a separate subsection of assets which can be translated as deferred charges (*Rechnungsabgrenzungsposten*). At the company's option, discounts on notes and bonds payable, and deferred tax assets may also be included in this category of assets.

Anlagevermögen (long-term assets) are defined as those assets that are intended to serve the operations of a company on a continuous basis (Lück 1983, 64). They represent the income producing assets of the company. Long-term assets are subclassified as intangible assets (*immaterielle Vermögensgegenstände*), fixed assets or property, plant and equipment (*Sachanlagen*), and financial assets (*Finanzanlagen*). A difference from United States asset classification is that land and buildings are combined on one line. A change from the previous law is that land and buildings are now reported on one line whereas previously they were required to be split into four separate categories—"factory land and building," "housing land and building," "land without buildings," and "buildings on non-owned land."

It can be seen from Table 2 that assets also include a category that can perhaps best be described as "suspense accounts"(*Bilanzierungshilfe*). This item is discussed below in the subsection on intangible assets.

Other differences in classification of assets between the United States and Germany are:

1. assets are shown in reverse order of liquidity,
2. treasury stock is included in marketable securities,
3. fixed assets includes all property, plant, and equipment including those not used in operations (e.g., land held for resale), and
4. inventories; property, plant, and equipment; and intangible assets each includes a category called "payments in advance."

For most German companies, the above mentioned differences from United States balance sheet classification are probably not material, and with required disclosure of the long-term portion of receivables and treasury stock, adjustments can be made to enhance comparability.

Equities

In the liability section of the balance sheet, German accounting practice distinguishes between actual legal obligations (*Verbindlichkeiten*) and provisions for estimated liabilities (*Rückstellungen*). The recording of estimated liabilities provides companies one of their greatest opportunities to create hidden reserves which is discussed below in the section on "Measurement of Liabilities." There is no distinction made on the face of the balance sheet between long-term and current liabilities. For example, item D.2. (Table 2), liabilities to financial institutions, can include both 90-day and 5-year notes payable. Companies are required, however, to disclose the amount of those obligations maturing within one year either parenthetically or in the notes.

The major differences from United States owners' equity classification relate to (1) the requirement that German companies must classify minority interest as a component of owners' equity and (2) the reporting of retained earnings.

There is no direct equivalent of an "unappropriated retained earnings" account in German financial reporting. How the current period's income and retained earnings are reflected in the financial statements depends on whether the annual report is published before or after the Board of Executive Officers has proposed the distribution of current profit. The differences can perhaps best be understood through the use of an example.

Assume that a company generates profit (*Jahresüberschuss*) of DM100 in 1990. There are three ways the board can propose to dispose of 1990 profit. It can be (1) paid out as dividends (*Ausschüttungen*), (2) appropriated to "other reserves" (Item A.III.4, Table 2) as retained earnings, and/or (3) carried forward (Item A.IV.) for the payment of dividends next year (*Gewinnvortrag*). Assume that the board proposes to pay out 40% of current profits in dividends, appropriate 50% to the reserve for retained earnings, and carry 10% forward to the following period. If financial statements are prepared after this proposal, disposition of the current period's earnings would appear in the financial statements as is shown in Table 3, panel A. As can be seen, the item "balance sheet profit" (*Bilanzgewinn*) reflects the amount earmarked for dividends. The "profit carried forward" (*Gewinnvortrag*) of DM10 would be added to 1991 net income in arriving at the amount of profit to be disposed of in 1991. "Other reserves" (*andere Gewinnrücklagen*) includes that portion of income that has been reinvested in the business.

Panel B of Table 3 shows the presentation of retained earnings assuming financial statements are prepared before the board's proposal for disposition of profits. In that scenario, net income is simply carried from the income statement to the owners' equity section of the balance sheet.

INCOME STATEMENT

Two income statement formats are now permissible in Germany; the total cost format (*Gesamtkostenverfahren*) and the cost-of-goods-sold format (*Umsatzkostenverfahren*) (see Table 4). The latter, introduced by the new law, results in the calculation of gross profit and is similar to the multiple-step format commonly used in the United States.

Table 3.   Presentation of Retained Earnings

A.   40% Dividend Payout, 50% Appropriated to Reserve for Retained Earnings

Income Statement

Jahresüberschuss	Net income	DM 10(
Einstellung in andere Gewinnrücklagen	Appropriation to other reserves	(50
Gewinnvortrag	Profit carried forward	(10
Bilanzgewinn	Balance sheet profit	DM 4(

Balance Sheet

Eigenkapital		Owners' equity	
III.	Gewinnrücklagen	III. Reserves	
	4. andere Gewinnrücklagen	4. other reserves	5(
IV.	Gewinnvortrag	IV. Profit carried forward	1(
V.	Bilanzgewinn	V. Balance sheet profit	4(
			DM 10(

B. No Disposition of Profit

Proposed Income Statement

Jahresüberschuss	Net income	DM 10(

Balance Sheet

	Eigenkapital	Owners' equity	
III.	Gewinnrücklagen	III. Reserves	
	4. andere Gewinnrücklagen	4. other reserves	0
IV.	Jahresüberschuss	IV. Net income	100
			DM 100

The total cost format is less familiar in the United States. Cost-of-goods-sold and general, selling and administrative expenses are not reported, instead aggregate amounts of materials cost, personnel cost, and depreciation are reported on separate lines. This was the only format allowed prior to 1986 and was the method prescribed in the draft Accounting Directives Law. Subsequent pressure exerted by industry, in particular those companies interested in following international accounting norms, caused the legislature to also allow the cost-of-goods-sold format. Sanction of the cost-of-goods-sold format continued to be debated in parliament until the final vote on the grounds that it diminished comparability across companies. As a compromise, the new Law requires that those companies using the cost-of-goods-sold format must disclose total materials costs and total personnel costs in the notes to the financial statements (HGB 285(8)).

Table 4. Income Statement Presentation

Total Cost (*Gesamtkosten*) format:	Cost-of-goods sold (*Umsatzkosten*) format:
1. Sales	1. Sales
2. Increases (decreases) in finished goods and work-in-progress	2. Cost-of-goods-sold
	3. Gross profit
3. Other capitalized self-constructed assets	4. Selling expenses
4. Other operating revenues	5. General administrative expenses
5. Materials costs	6. Other operating revenues
a. raw materials and supplies	7. Other operating expenses
b. expenditures for services	
6. Personnel costs	OPERATING PROFIT
a. wages and salaries	
b. social security and pensions, of which for pensions	
7. Depreciation	
a. on intangible assets, fixed assets, and capitalized organization costs	
b. write-downs of current assets, to the extent these exceed normal amounts	
8. Other operating expenses	
OPERATING PROFIT	

The remaining items are common to both formats:
9. Income from stock investments, of which from subsidiaries
10. Income from other marketable securities and loans, of which from subsidiaries
11. Other interest and similar income, of which from subsidiaries
12. Write-downs of financial assets and current marketable securities
13. Interest and similar expenses, of which to subsidiaries
14. INCOME (LOSS) FROM OPERATIONS
15. Extraordinary income
16. Extraordinary expenses
17. Net extraordinary income (expense)
18. Income taxes
19. Other taxes
20. NET INCOME (LOSS)

Source: *Handelsgesetzbuch* Section 275.

Unusual Items Provided Special Treatment in the United States

Three items are provided special treatment in United States income statements: (1) discontinued operations, (2) extraordinary items, and (3) the cumulative effect of a change in accounting principle. Each of these is shown below "after tax income from continuing operations" and is therefore shown on a net of tax basis (intraperiod tax allocation). Of these three, only extraordinary items are provided a separate heading under German law and these are shown before provision for income taxes. Thus, intraperiod tax allocation is unknown in Germany. Extraordinary items are not as restrictively defined as in the United States. The law simply defines these as gains and losses that arise outside of the normal operations of the company (HGB Sec. 277(4)). Practice indicates that this heading would include extraordinary items as defined in the United States, as well as gains (losses) on disposal of operations and on sales of major assets.

The cumulative effect of a change in accounting principle as an adjustment to income is unknown in Germany. The law indicates that justified inconsistencies in accounting principle and their effect on net worth, financial position, and results be disclosed in the notes (HGB 284(2)3.) There is some question as to whether the effect must be quantified. In contrast to the United States, earnings per share information is not required and generally is not disclosed by German companies.

SCHEDULE OF CHANGES IN LONG-TERM ASSETS

In addition to a balance sheet and income statement, German companies are required to present a statement summarizing the increases and decreases in each of the items of long-term assets. As can be seen from Table 5 this schedule provides detailed information not commonly provided in the annual reports of United States companies. Under previous law, this schedule reported the change in book value during the period with no disclosure of original costs. The new law now requires companies to report original cost and accumulated depreciation separately. This represents one of the most important changes in the financial reporting of assets brought about by the new law. Coenenberg (1986, 73) suggests that the cost to adjust corporate financial reporting systems to accommodate the preparation of this schedule was substantial.

NOTES TO THE FINANCIAL STATEMENTS

Prior to 1986 there was no specific requirement that German companies prepare supplementary notes to the financial statements. Instead, companies were required to provide certain supplemental information in the management report. The innovation of the new law is that notes have become a third component of a set of financial statements (*Jahresabschluss*) along with the balance sheet and income statement. Coenenberg (1986, 167) suggests that the most important consequence of this is that the notes can be used to relieve the balance sheet and income statement from an overload of information.

As in the United States, German companies are required to provide a summary of significant accounting policies in the notes. The law specifically requires disclosure of foreign currency translation rules due to the fact that the law does not prescribe any particular rules to be followed.

Table 5. Schedule of Long-Term Assets

LONG-TERM ASSETS	COST				ACCUMULATED DEPRECIATION					BOOK VALUES	
	1Jan1989	Additions	Disposals	31Dec1989	1Jan1989	Additions	Disposals	Revaluations	31Dec1989	1Jan1989	31Dec1989
Patents, trademarks, and other legal rights											
Goodwill											
Intangible Assets											
Land and buildings											
Machinery and equipment											
Other fixed assets											
Payments in advance and construction in progress											
Fixed Assets											
Investments in subsidiaries											
Loans to subsidiaries											
Investments in associated entities											
Loans to associated entities											
Long-term marketable securities											
Other loans											
Financial Assets											
Total Long-Term Assets											

89

The bulk of the notes to be found in German annual reports provide additional information on individual balance sheet and income statement line items. For example, the new law requires German companies to disclose the amount of liabilities maturing in less than one year and in more than five years. This can be done parenthetically on the face of the balance sheet or, as is more common, in the notes. It is not uncommon for German companies to reference twenty or more notes on the face of their financial statements.

In addition to more detailed information on financial statement items, German companies are also required to present supplementary information in the notes. These include such items as:

- segment revenues, by industry and geographic market;
- contingent liabilities and financial commitments;
- average number of employees by category;
- remuneration and loans to members of the Supervisory Board and Board of Executive Officers; and
- name, location, ownership percentage, equity, and last year's profit for those shareholdings in which the company has a 20% or more ownership interest.

The last item usually takes the form of a one- or two-page Table often placed before the financial statements. United States companies typically do not provide information in their annual reports on major shareholdings in the level of detail required in Germany. It is interesting to note that the law provides an escape clause with regard to the disclosure of significant shareholdings. This information need not be reported if "under reasonable business judgment, it would result in a great disadvantage to either the investor or investee company" (HGB 286(3)2). This author is unaware of any major company in Germany availing itself of this escape clause.

MEASUREMENT AND VALUATION OF ASSETS

Several differences exist between United States GAAP and German law with regard to the valuation of assets. These primarily relate to the definition of historical cost, unplanned depreciation and asset write-downs, and capitalization versus expensing of costs.

Historical Cost

The Accounting Directives Law provides that assets are not to be stated at more than their purchase or manufacturing cost reduced by sanctioned write-downs. Purchase cost is defined similarly as in the United States, that is, expenditures incurred to acquire an asset and place it in usable condition. There is no specific definition of how manufacturing cost is to be determined. A lower limit is prescribed and companies are then provided options as to whether certain costs are capitalized or expensed immediately.

Inventory Cost

The minimum value allowed by law for determining the cost of inventories is direct or prime cost. Allocation of manufacturing overhead to product is allowed, but optional. The law also allows allocation of general administration costs and, in certain situations, interest costs to inventory. Thus, the alternative acceptable measures of inventory cost are:

1. direct cost,
2. direct cost plus variable overhead,
3. direct cost plus variable and fixed overhead (full cost),
4. full cost plus general administration costs, and
5. full cost plus general adr..inistration costs plus interest costs.

In contrast, in the United States, manufacturing overhead must be allocated and interest costs may not be allocated to inventory. By choice, German companies could determine the cost of inventory in a manner consistent with United States rules. An examination of inventory valuation principles of a number of German companies found that a majority have opted to allocate overhead, but not general administrative and interest costs. This is probably a result of the fact that German tax law requires inventory to be carried at full cost.

Write-downs of Assets

As mentioned above, one of the major sources of hidden reserves for German companies is the write-down of assets for reasons other than normal depreciation. Current assets must be carried at the lower-of-cost-or-market. However, current assets must be further written down below current market value if in the exercise of sound business judgment (*vernünftiger kaufmännischer Beurteilung*) their value is likely to fall even further in the near future. Long-term depreciable assets (intangibles and fixed assets) must be amortized according to some plan. Unplanned, extraordinary write-downs (*ausserplanmässige Abschreibungen*) may be taken on all long-term assets (including investments) if a future permanent reduction in value is foreseeable. Furthermore, the new law indicates that write-downs of both long-term and current assets are to be taken if, in the exercise of sound business judgment, they are deemed necessary. Thus, German companies are provided considerable freedom in writing down the values of their assets consistent with the emphasis on conservatism, and practice suggests that they freely avail themselves of this opportunity.

Under previous law, companies were not required to reverse write-downs of assets if the reason for the write-down no longer exists (e.g., if the prognosis for lower values does not materialize). The new law has somewhat reduced companies' ability to maintain hidden reserves with regard to asset write-downs. In compliance with the Fourth Directive, companies are generally required to revalue their assets when the reason for the write-down no longer exists. However, to achieve tax neutrality (a major objective of the legislature in transforming the Fourth Directive into national law), German companies are not required to reverse previous writedowns if this would result in an increase in tax liability. Thus, the new law only requires reversal of write-downs in situations where the *Massgeblichkeitsprinzip* does not hold. This is the case with respect to "write-ups" of depreciable long-term assets (intangibles and fixed assets). In other words, because the *Massgeblichkeitsprinzip* is in force with regard to revaluations of current assets and long-term investments, the Accounting Directives Law allows write-downs on these categories of assets to be maintained even if they are no longer justified.

Inflation Accounting

The EC's Fourth Directive gave member nations the option to require their companies to adjust their accounts for inflation. In the past, the German government has been opposed to inflation accounting because of its alleged detrimental effect as a stimulus to future inflation (Macharzina 1981, 145). Through the requirement that assets not be carried at more than their purchase or manufacturing cost, the Accounting Directives Law implicitly disallows the use of inflation accounting.

Interest Capitalization

The Accounting Directives Law indicates that interest on borrowed capital used to finance the production of an asset may be capitalized to the extent that it is incurred during the period of production (HGB 255(3)). This applies equally to the production of inventories and fixed assets. Literal interpretation of the law would imply that interest may only be capitalized when funds have been borrowed directly to finance construction of an asset. Thus, capitalization of an amount of so-called "avoidable interest" in accordance with SFAS 34 would appear not to be acceptable. Given the emphasis on the conservatism principle, German companies generally do not exercise the option to capitalize interest.

Valuation Adjustment Account for Tax-Related Items
(Sonderposten mit Rücklageanteil)

As mentioned earlier, write-downs of both current and long-term assets are generally more liberal in Germany than in the United States. Section 254 of the Accounting Directives Law specifically states that write-downs may be recorded to state fixed and current assets at the lower value which results from applying special tax-allowed write-downs.

To the extent that tax law requires that special tax-allowed depreciation be reflected in the financial statements (the reverse *Massgeblichkeitsprinzip*), these amounts may be shown in the balance sheet in a valuation adjustment account which may be translated as "special account with a reserve component" (*Sonderposten mit Rücklageanteil*) placed between owners' equity and estimated liabilities. For example, to stimulate investment in the city of Berlin, Section 14 of the Berlin Investment Incentive Act (BFG) allows companies to take 75% depreciation in the first year on fixed assets located in Berlin. Assume that a building is constructed at a cost of DM100 million and will be depreciated on a straight-line basis over 20 years. In year one, the DM70 million difference between tax depreciation of DM75 million and book depreciation of DM5 million would be credited to a "special account" shown on the equities side of the balance sheet. Through this device, the building is carried at year-end at DM95 million which more closely approximates its "true and fair" value at that time than would DM25 million, and the company has complied with tax law by reflecting tax depreciation on the balance sheet. Because tax depreciation of DM75 million is taken in calculating accounting income, no provision for deferred taxes is required. In fact, this "special account" can be analyzed as consisting of two components: 50% deferred tax liability and 50% retained earnings. This can be seen through a continuation of the above example.

Income Statement (with 75% depreciation]			*Income Statement* (with 5% depreciation)		
Sales		125	Sales		125
Dep.		75	Dep.		5
Inc. before taxes		50	Inc. before taxes		120
Income taxes 50%		25	Income taxes 50%		60
Net income		25	Net income		60

Balance Sheet (with 75% depreciation)				*Balance Sheet* (with 5% depreciation)			
Building	100	Capital	100	Building	100	Capital	100
Acc.dep.	(5)	Spec.Acct.	70	Acc.dep.	(5)	Def.Tax	35
Cash	100	Ret. earn.	25	Cash	100	Ret. earn.	60
Total	195		195	Total	195		195

Given a German corporate tax rate of 50%, the special account of 70 decomposes into a deferred tax liability of 35 and retained earnings of 35.

The reverse *Massgeblichkeitsprinzip* only applies to parent company statements. Thus, for consolidated financial statements, the "special account" can be dissolved with correspondingly higher deferred taxes and retained earnings. For example, in 1989, Schering AG reported deferred taxes of DM15 million and a special account of DM484 million in the parent company only statements. In the consolidated statements, deferred taxes were DM278 million and the special account was dissolved.

It must be noted that the use of the "special account" is optional. Companies are also allowed to carry the asset at the lower value allowed for tax purposes. In the above example, Building could have been carried on the balance sheet at DM25 million at the end of year 1.

Intangible Assets

Just as in the United States, the Accounting Directives Law prohibits capitalization of research and development costs. The law also prohibits capitalization of costs incurred in obtaining internally developed intangible assets such as patents, trademarks, and goodwill. These may only be recorded as assets when purchased from outside parties. United States accounting rules require that the direct costs of obtaining such legal rights be capitalized and amortized over the lower of useful and legal life. However, as these costs are insignificant in many instances, application of the materiality constraint allows companies in the United States to expense these costs immediately.

Organization costs are not considered assets in Germany. Due to the emphasis on creditor protection, German accounting has traditionally followed the principle that assets must be separably marketable items. If an item cannot be separated from the business and sold, then it cannot be used to generate cash to pay creditors, and it is not an asset. This definition conflicts with the concept of cost deferral for a better matching of revenue and expense which is the purpose for capitalizing organization costs in the United States. Moreover, as pointed out by Coenenberg (1986, 63), immediate expensing of organization costs exacerbates the probable loss generated by companies in the first year of operations. To overcome this dilemma, the new law created a suspense account entitled *Bilanzierungshilfe* in which companies are allowed at their discretion to defer the costs of starting-up and expanding the business (*Aufwendungen für die Ingangsetzung and Erweiterung des Geschäftsbetriebes*). Although technically not an asset, this account would appear as the first item on the asset side of the balance sheet. A similar option is available for deferred income tax assets except that this would be reported as a deferred charge (*Rechnungsabgrenzungsposten*) at the bottom of the asset side of the balance sheet. The law further requires that if organization costs or deferred tax assets are deferred in the balance sheet, profits may be distributed to stockholders only to the extent that the reserve for retained earnings exceeds those amounts. Capitalized organization costs must be amortized at a rate of at least 25% per year.

Purchased goodwill may be either capitalized and amortized or written-off directly to owners' equity. The law states that if capitalized, goodwill must be amortized over four years or, alternatively, over the number of years that are likely to benefit. Tax law allows goodwill to be written off over 15 years and this period is therefore commonly chosen for accounting purposes as well.

Capitalized goodwill is classified as an intangible asset in the balance sheet. Because goodwill cannot be separately sold, it does not meet the German definition of an asset. To be consistent, the legislature should have required that goodwill be handled in a similar manner as capitalized organization costs.

Negative goodwill must be set up on the equities side of the balance sheet and may be taken to income only if:

1. an unfavorable development in the results of the enterprise which was anticipated at the time of acquiring the shares has occurred or expenditures anticipated at that time must be recognized, or
2. it becomes clear on the balance sheet date that it corresponds to a realized profit (HGB Sec. 309(2)).

Thus, in contrast to the practice of systematic amortization in the United States, negative goodwill is deferred indefinitely and then written off as a lump sum when certain conditions arise.

MEASUREMENT OF LIABILITIES

The general rule pertaining to the measurement of liabilities provides that liabilities are to be stated at their maturity amount, pension obligations at their present value, and estimated liabilities at the amount required based on sound business judgment (HGB 253(1)).

Estimated Liabilities

Section 249 of the HGB governs the accounting for estimated liabilities (*Rückstellungen*). There are basically two types of estimated liabilities that German companies are required or allowed to accrue: loss contingencies and future expenditures. As under previous law, the Accounting Directives Law requires German companies to set up provisions for "uncertain liabilities" (*ungewisse Verbindlichkeiten*) and "anticipated losses from uncompleted transactions" (*drohende Verluste aus schwebenden Geschäften*). There are no explicit criteria in the law (such as those of "probable" and "reasonably estimable" under SFAS 5 in the United States) to guide companies as to when provisions for uncertain liabilities must be made. Thus, German companies have freer rein in accruing contingent liabilities and losses than their United States counterparts. An innovation in the new law is that German companies are now expressly required to accrue estimated warranty obligations.

The new law also requires companies to estimate a liability and establish a provision for (1) postponed maintenance and repairs to be made in the first three months of the following period and (2) land reclamation expenses to be incurred in the following year. Under previous law, accrual of these items was optional. The new law also gives companies an option to accrue a liability and expense for postponed maintenance and repairs to be made after three months but within the next twelve months. There is no direct equivalent of these requirements in the United States.

The greatest innovation in the new law regarding estimated liabilities is the option to create provisions for future expenditures under certain conditions:

1. the nature of the future expenditure is specifically identified,
2. the reason for the future expenditure results from conditions existing at the balance sheet date,
3. it is probable or certain that the expenditure will be made, and
4. the magnitude or date of expenditure is unknown.

Coenenberg (1986, 107) claims that the German accounting literature views the sanction of provisions for future expenditures with great skepticism. He suggests that the ability to make such provisions greatly increases management's "room to play with the financial statements" (*bilanzpolitischer Spielraum*), so much so that this could become a "provision for threatening profits" (*Rückstellung für drohende Gewinne*). In other words, if profits are greater than the company wishes to show, then these may be reduced by a provision for future expenditures. The following example illustrates how this would work.

In December 19X1, management determines that major repairs to an assembly line estimated to cost DM2,000,000 will have to be made within the next several years in order for operations to continue. In 19X1, the company accrues a liability and reports "other operating expenses" of DM2,000,000. Thus, the company is able to reduce profit in 19X1 for an expenditure that may not be made until several years in the future. It should be noted, however, that this expense would not be recognized for tax purposes.

As an example of the magnitude estimated liabilities can attain in German companies, Schering AG reported consolidated sundry provisions in 1989 for "bonuses, early retirement costs, sundry personnel costs, environmental protection measures, maintenance, price allowances, exchange rate fluctuations and other risks in foreign trading, as well as other uncertain liabilities." At year-end 1989 the accumulated sundry accruals amounted to

12% of total assets and the increase in this account amounted to 45% of consolidated pre-tax profit.

DEFERRED TAXES

There was no express requirement for deferred tax accounting in Germany under previous law. However, many companies reported deferred tax liabilities as a matter of practice. The EC's Fourth Directive requires both deferred tax assets and liabilities to be recognized, either in the balance sheet or in the notes to the financial statements. In implementing the Fourth Directive, the German legislature was therefore required to introduce mandatory deferred tax accounting into German accounting practice.

The Accounting Directives Law requires that deferred tax liabilities be reported either in the balance sheet as an estimated liability or in the notes. On the other hand, deferred tax assets may be reported in the balance sheet as a deferred charge, but they need not be, and if not, there is no requirement that they be disclosed in the notes. It is unclear whether this was intentional or merely an oversight on the part of the law's authors.

Timing differences giving rise to deferred tax liabilities may be netted with those giving rise to deferred tax assets to determine a net deferred tax asset or liability for the period, with an option as to whether a net deferred tax asset would be disclosed. The new law also allows deferred tax assets and liabilities to be determined and reported separately. In that case, however, the deferred tax asset would be required to be reported in the balance sheet.

The law does not specify a method (deferral or liability) of tax effect accounting: both appear to be acceptable. Moreover, the conservatism principle in Germany would appear to allow companies to use a modified liability method, adjusting for changes in tax rates only when this would generate an increase in a deferred tax liability or a decrease in a deferred tax asset (Coenenberg 1986, 129).

PENSION ACCOUNTING

One of the most important innovations introduced by the Accounting Directives Law was the obligatory accrual of pension obligations. The spectacular bankruptcy of the German company AEG (General Electric) showed the German business community how dangerous it can be to allow companies to ignore their pension obligations in their financial statements. German companies do not externally fund their pension plans. If the company goes bankrupt, retired employees are left stranded. In the case of AEG, the pension obligation was subsequently funded by the rest of German industry. In writing the Accounting Directives Law, the legislature hoped that the obligatory accrual of pension obligations will help to avoid another AEG debacle.

The Implementation Law of 1987 includes transitional provisions for the Accounting Directives Law. Article 28 of that law provides that a provison need not be established for pension obligations existing prior to January 1, 1987, thereby allowing estimated liabilities to be understated. This significantly reduces the protection provided retirees and is unusual in German accounting practice in that it is inconsistent with the conservatism principle. However, the damage is at least partially repaired in that companies must report the amount not accrued in the notes to the financial statements. Other than this, the new law does not require any specific disclosures regarding pensions to be made in the notes.

Due to the fact that German companies do not externally fund their pension obligations, the financial reporting of pensions in Germany is considerably different from that in the United States. Whereas in the United States the amount of pension obligation or pension asset reported in the balance sheet is equal to the difference between accumulated benefit obligation and fair value of pension plan assets, in Germany, since no pension plan assets exist per se, the obligation reported on the balance sheet is equal to the accumulated benefit obligation. Thus, compared to United States companies, the current period's provision for pension obligations for German companies generally constitutes a higher percentage of total equities. Because pension plans are not externally funded, the accumulated provison for pension liability represents a significant source of financing for many companies. In a sense, the pension plan assets are invested internally in fixed assets, inventories, and so on. Economically, the tax shield provided by the pension expense, the interest cost avoided through internal financing, and the internal return on investment can combine to substantially cover the cash outflow of a company's pension plan.

Regarding valuation, the new law simply requires that pension liability be stated at its present value. For guidance as to how to calculate the present value of pension obligations, German companies must refer to the Tax Law which requires the use of a "partial value method" (*Teilwertverfahren*) in calculating current pension expense. This method differs from that required in the United States under SFAS 87. In the United States, the calculation of pension obligation is based on years of service rendered to date and estimated future salary levels when employees retire (projected benefits obligation method). German tax law requires pension expense to be calculated using current salary levels; a second situation in which United States practice is actually more conservative.

Another difference between United States and German pension accounting relates to the treatment of prior service costs. In the United States these costs must be capitalized and amortized over the average remaining service life of employees. In Germany, these costs must be either charged to income immediately or capitalized and amortized over three years. If capitalized, they would properly be classified as a *Bilanzierungshilfe*, that is, in the suspense account.

LEASE ACCOUNTING

The Accounting Directives Law is silent with regard to the accounting for leases. The new law requires only that "other financial commitments" be disclosed in the notes which would appear to include lease commitments.

The capitalization of leases is supported in German accounting literature when the lease contract is in reality an installment purchase (see, for example, Adler/Düring/Schmaltz 1987, Section 149), and the new law does not preclude companies from adopting this policy. As a result, German companies are free to choose their own policy and develop their own criteria for when leases should be capitalized. For example, one company has adopted the policy that leases should be capitalized only if at the end of the lease period transfer of property takes place or a purchase option exists which is required by contract to be exercised. For this company, lease capitalization criteria are significantly more restrictive than those of SFAS 13.

CONSOLIDATION

In integrating the EC's Seventh Directive into German accounting practice, the German legislature introduced significant changes in both the scope and preparation of consolidated financial statements. A significant innovation in the new law requires consolidated financial statements to be prepared as if all the companies consolidated comprise one entity. This one-entity theory (*Einheitstheorie*) underlies the uniform valuation requirement introduced by the new law which requires that the assets and liabilities of the enterprises included in the consolidated financial statements be valued uniformly in accordance with the valuation methods applied in the financial statements of the parent (HGB Sec. 308). This contrasts with previous law which required that amounts carried in the subsidiaries' financial statements be consolidated without any conversion to a common set of valuation principles.

Consolidation must now be effected following what the Germans call the Anglo-Saxon method (*angelsächsische Methode*). Two options are allowed which equate to the parent-company and entity concepts of consolidation (see Jensen, Coffman, and Burns 1988, 80-2). Under either option, the net assets of the acquired company are revalued at the date of acquisition or first consolidation. The amount of revaluation and any goodwill is fixed at that time. Under previous law there was no revaluation of net assets. The parent's investment was eliminated against the subsidiary's book value of net assets at each balance sheet date with a different amount of "consolidation difference" arising each time. Elimination of losses on intercompany asset sales, which was not previously required, is also now required under the new law. Thus, significant changes have been introduced by the new law with regard to consolidation procedures.

The Accounting Directives Law requires German companies for the first time to consolidate their foreign subsidiaries. With this innovation, the scope of consolidation in Germany corresponds with that in the United States with one major exception. Companies with activities that diverge from the rest of the group (e.g., finance subsidiaries), must be excluded from consolidation. Beginning in 1988, in accordance with SFAS 94, U.S. companies have been required to include such subsidiaries in the consolidated financial statements. There is no requirement in Germany that separate financial statements of nonconsolidated subsidiaries be presented in the annual report as was required in the United States prior to SFAS 94.

The new law requires consolidation of foreign subsidiaries but provides no guidance as to how the foreign currency financial statements should be translated into deutsche marks. Similar to SFAS 1, the law only requires disclosure of the method used. A variety of methods is used by companies in Germany; primarily the monetary-nonmonetary or temporal method and the current (or closing) rate method.

Some companies have adopted SFAS 52's functional currency approach or IASC 21's integral operations approach in selecting the appropriate translation methodology for individual foreign operations. In contrast with United States parent companies who predominately view their foreign operations as relatively independent from the parent and therefore use the current rate method of translation with deferral of translation gains and losses, German companies tend to view their foreign operations as integral to the operations of the parent and therefore prefer the use of the temporal method with translation gains and losses taken to income. Given that dramatically different balance sheet and income statement amounts can result from application of different translation approaches, the free choice of method in Germany could be the greatest source of noncomparability of consolidated financial statements, not only within Germany, but also between Germany and the United States. On the other hand, German companies interested in preparing financial statements comparable to those in the United States are provided the option within German law to apply SFAS 52.

Another significant change brought about by the new law relates to the use of the equity method. For the first time, German companies are now required to use the equity method for investments in which the investor has a 20% or greater ownership interest.

SEGMENTAL REPORTING

Segmental reporting is a disclosure issue in which United States and German accounting practices are significantly different. The Accounting Directives Law requires disclosure in the notes to the financial statements of sales by area of activity as well as by geographically defined markets "to the extent that the areas of activity and geographically defined markets differ significantly from each other, taking into account the sales organization for the goods and services typical of the company's ordinary business" (HGB Sec. 314(1)). Major differences from United States practice under SFAS 14 include:

1. no required disclosure of profit and assets by segment,
2. no minimum threshold for when disclosure must be made,
3. geographic segment defined as market rather than area, and
4. emphasis on sales organization in defining segments.

Although the new law does not prohibit German companies from reporting profit and assets on a segmental basis, for competitive reasons, few, if any, German companies voluntarily provide this information. Moreover, the law also provides companies with an escape clause. Segment sales need not be reported if it is expected that the disclosure will cause material harm to an enterprise included in the consolidated financial statements (HGB Sec. 313(2)).

SUMMARY AND CONCLUSIONS

The purpose of this paper was to describe the changes in German accounting practice brought about by the Accounting Directives Law of 1985 and compare current German accounting practice with that of the United States.

With the integration of the EC Directives into German law, readers of financial statements knowledgeable of United States practices should find German annual reports easier to read and understand than in the past. This is especially true for those companies opting to use the newly allowed cost-of-goods-sold format income statement.

With regard to disclosure, readers will now find a familiar looking set of notes with information on balance sheet and income statement items often more detailed than is found in the United States. The only disclosure areas in which Germany still lags behind the United States are segmental reporting and cash flows. Whereas major German companies provide cash flow information on a voluntary basis in an attempt to conform to international norms, this is not the case with regard to reporting segmental profits.

Whether the amount of profit reported in Germany is consistent with what would be reported under United States rules greatly depends on the extent to which German companies avail themselves of the opportunity to create hidden reserves. There are actually few areas in which German law *requires* different treatment from that in the United States. Other than these items, German companies that are not concerned with income smoothing or income tax minimization could conceivably comply with United States accounting rules in all material respects. In fact, a movement in this direction may already be underway. In conjunction with listing its stock on the London Stock Exchange, Schering AG has adopted a policy of complying with internationally recognized norms of accounting to the extent allowed by German law. In an insert placed in its 1989 Annual Report to British stockholders, the company reported that it complies with IASC standards in all material respects other than nonrecognition of unrealized foreign exchange gains and nonreporting of segmental profits. Although IASC standards and United States GAAP are not identical, differences are not substantial. Daimler-Benz's uncovering of hidden reserves in 1989 provides additional anecdotal evidence that German companies may be interested in complying with international norms of accounting.

REFERENCES

Adler/Düring/Schmaltz. 1987. *Rechnungslegung und Prüfung der Aktiengesellschaft*, 5. Auflage, prepared by K. Schmaltz, K-H. Forster, R. Goerdeler, and H. Havermann. Stuttgart: C.E. Poeschel Verlag.

American Accounting Association. 1977. Report of the 1980-1981 International Accounting and Auditing Standards Committee. *The Accounting Review* (Supplement): 65-101.

Choi, F., and V. Bavishi. 1982. Diversity in multinational accounting. *Financial Executive* (August): 45-49.

Coenenberg, A. 1986. *Die Einzelbilanz nach neuem Handelsrecht.* Düsseldorf: Industriebank AG.

Der Spiegel. 1990. Daimler Benz—Wundersame Vermehrung. April 23, pp. 114-16.

Forbes. 1990. The long ascent from stunde null. July 23, pp. 122-23.

Heinen, E. 1980. *Handelsbilanzen*, 9. Auflage. Wiesbaden: Gabler.

Imhoff, E. 1981. Income smoothing: An analysis of critical issues. *Quarterly Review of Economics and Business* (Autumn): 23-42.

Jensen, D., E. Coffman, and T. Burns. 1988. *Advanced Accounting.* 2nd ed. New York: Random House.

Jung, A, H. Isele, and C. Gross. 1989. Rechnungslegung in den USA. In *Rechnungslegung, -prüfung, Wirtschaftsrecht, und Steuern in den USA*, edited by E. Sonnemann. Düsseldorf: Gabler.

Lück, W., ed. 1983. *Lexikon der Betriebswirtschaft.* Landsberg am Lech: Verlag Moderne Industrie.

Macharzina, K. 1981. Financial reporting in West Germany. In *Comparative International Accounting*, edited by C. Nobes and R. Parker. Homewood, IL: Irwin.

Mueller, G. 1968. Accounting principles generally accepted in the United States versus those generally accepted elsewhere. *International Journal of Accounting Education and Research* (Spring): 91-103.

Nair, R., and W. Frank. 1980. The impact of disclosure and measurement practices on international accounting classifications. *The Accounting Review* (July): 426-50.

Nobes, C., and R. Parker, eds. 1981. *Comparative International Accounting*. Homewood, IL: Irwin.

ACCOUNTING AND AUDITING STANDARDS OF CANADIAN DEVELOPMENT STAGE COMPANIES OPERATING IN THE U.S. INVESTMENT MARKET

Michael L. Costigan and Norlin G. Rueschhoff

ABSTRACT

This paper reports a study of differences in the way that Canadian development stage firms decide among accounting and auditing alternatives for reporting in the U.S. securities market. The reporting decisions are particularly significant because of U.S./Canadian conflicts in accounting for general and administrative costs in the start-up stage and in reporting uncertainties and going concern considerations in the early stages of development. Under U.S. Securities and

Advances in International Accounting,
Volume 5, pages 105-114.
Copyright © 1992 by JAI Press Inc.
All rights of reproduction in any form reserved.
ISBN: 1-55938-415-8

Exchange regulations, Canadian firms do have alternatives for reporting under conflicting situations. The study not only reports on the conflicting accounting principles but also summarizes and analyzes the accounting and auditing bases used by the firms in their reporting decisions.

INTRODUCTION

Recently, members of the American and Canadian accounting professions have commenced negotiations to permit reciprocity between the American Certified Public Accountant certificate and the Canadian Chartered Accountant certificate. This comes at a time when considerable international financial and investment interchange exists between the two countries even though significant differences continue to exist between U.S. and Canadian accounting and auditing principles. For example, as this study reports, over one-half of the almost 500 foreign firms with common shares actively traded in the U.S. securities market are Canadian firms. Some are rather mature firms with shares listed on either the New York Stock Exchange or the American Stock Exchange whereas others are in their development stage, operating in the U.S. securities over-the-counter market.

The Canadian firms in their development stage must consider special accounting and auditing standards when they decide to operate in the U.S. securities market. The particular standards related to accounting and auditing across U.S./Canadian borders must be considered. If reciprocity between the two countries is to be achieved, harmonization of accounting and auditing standards for companies seeking equity financing across borders is quite important. In light of this situation, the relevant accounting and auditing standards are first reviewed. Then the methodology of the study is presented along with a description of the sample selection. The results of the research are next reported, followed by an analysis of the reporting decision alternatives according to the alternative accounting and auditing bases. Finally, a summary and conclusions are presented.

ACCOUNTING FOR DEVELOPMENT STAGE ENTERPRISES IN THE UNITED STATES

The financial reporting features of development stage enterprises were studied in 1973, with illustrated financial statements presented for such companies (Crooch and Collier 1973). This was before the Financial Accounting Standards Board issued FASB Statement No. 7, *Accounting and Reporting by Development Stage Enterprises* (FAS7) in June 1975. FAS7 specified the guidelines for identifying an enterprise in the development stage as well as the standards of financial accounting and reporting applicable to such enterprises. In accordance with FAS7, a company is a development stage enterprise when it is devoting most of its efforts to activities such as financial planning, raising capital, exploring for natural resources, developing natural resources, research and development, establishing sources of supply, acquiring plant, property, and equipment, acquiring other operating assets such a mineral rights, recruiting and training personnel, developing markets, and starting up production. Though the development stage company advances into an operating stage when planned principal operations have commenced, it continues to be a development stage enterprise until significant revenues accrue. In 1984, illustrations of accounting and reporting by development stage enterprises were published (Goodman and Lorensen 1984).

In the United States, financial statements for development stage enterprises must conform to generally accepted accounting principles applicable to established operating enterprises. Further, under FAS7, the financial statements must also provide additional disclosures, including reports of any cumulative net losses in the stockholders' equity section of the balance sheet, cumulative amounts of revenue and expenses from the enterprise's inception in the income statement, cumulative amounts of sources and uses of financial resources in the cash flow statement, and detailed disclosure of each share issuance in the statement of stockholders' equity. The financial statements must identify the firm as a development stage enterprise and provide a description of the nature of the development stage activities.

UNITED STATES/CANADIAN CONFLICTS IN REPORTING

To date, Canadian accounting principles have not dealt with the topic of accounting by development stage enterprises. However, from a recent comparative study reporting major differences between United States and Canadian accounting principles (Most 1984, 52-53), one such conflict is reported for development stage enterprises. The Canadian principles allow development stage firms to defer general and administrative costs during the development stage while FAS7 does not allow such deferrals.

Further, a Canadian Auditing Guideline was issued in 1981 and revised in 1988 to deal with United States-Canadian conflicts in reporting (Canadian Institute of Chartered Accountants 1988). The guideline deals with conflicts in disclosure with respect to contingencies and going concern considerations for companies operating in the United States. Such conflicts are a significant disclosure issue for development stage companies because excessive uncertainties often exist for these companies and cause a concern as to their going-concern continuity. In Canada, as long as the uncertainties are fully disclosed in the financial statements, no audit report qualification nor further explanation by the auditors is needed. In the United States, however, such uncertainties must be fully disclosed in the auditor's report. The conflict in disclosure requirements is resolved by the Canadian profession through the issuance of a special supplementary report entitled "Comments by Auditor for U.S. Readers on Canadian-U.S. Reporting Conflict." The suggested wording for the supplementary report is as follows (Canadian Institute of Chartered Accountants 1988, 5):

> In the United States, reporting standards for auditors require the addition of an explanatory paragraph (following the opinion paragraph) when the financial statements are affected by significant uncertainties such as those referred to in the attached balance sheets as at December 31, 19x2 and 19x1 and as described in Note…of the financial statements. My report to the shareholders dated …………, 19.. is expressed in accordance with Canadian reporting standards which do not permit a reference to such uncertainties in the auditor's report when the uncertainties are adequately disclosed in the financial statements.

Finally, under United States Securities regulations, accounting principles generally accepted in either the United States or Canada may be used. If Canadian accounting principles are used, a reconciliation to U.S. generally accepted accounting principles must be disclosed as part of the financial statements (U.S. Securities and Exchange Commission 1988, par. 210.4-01(a)(2)).

THE METHODOLOGY FOR THIS STUDY

In deriving the sample of firms, the Compact Disclosure data base was used. All foreign firms reporting as development stage companies in the data base were identified. Of 486 foreign companies with fiscal years from 1986 through 1988, 34 different companies were disclosed as Canadian development stage companies. Twenty-six of the 34 Canadian firms were involved in natural resource exploration and mining at the nonoperating stage. The report of one of these Canadian nonoperating mining companies was unavailable. Thus, this study includes 33 Canadian development stage enterprises operating in the U.S. investment market. Twenty-five are companies in the nonoperating development stage. The other eight are operating companies that have reported revenue from sales. A list of these companies is presented in the Appendix. The data were obtained from the corporate annual reports.

The research consists of two steps. First, the differences in accounting standards that were reported in the annual reports were summarized. Then, the accounting and auditing bases used by the firms in their reporting decisions were analyzed. Specifically, the elements of the accounting decisional base include (1) the comprehensive body of accounting principles used in preparing the financial statements and (2) the reporting currency selected. The elements of the auditing decision base include (1) the audit standards applied and (2) the place where the audit report is issued.

Because the enterprises involved are subject to U.S. securities regulation, the financial statements must disclose the accounting principles used in preparation of the statements. Further, the reporting currency should be identified in the financial statements because the U.S. Securities and Exchange Commission regulations permit reporting in either Canadian dollars or U.S. dollars (U.S. Securities and Exchange Commission, 1988, pars. 210.3-20, 210.4-01(a)(2)). Canadian firms can justify the use of the U.S. dollar as a reporting currency if the U.S. dollar is the dominant functional currency (Rueschhoff 1973). A number of Canadian firms with shares listed on the New York Stock Exchange have used U.S. dollar reporting.

Further, the problems of being in the development stage and also subject to U.S. securities regulation will be apparent in the audit reporting decisions. An audit report may be issued under U.S. audit standards or under the audit standards of Canada, although the "comments to the U.S. readers" report may be added by the auditors. The country in which the audit report was issued is also of interest in viewing consistency in the audit reporting. In this regard, the International Auditing Guideline on audit reporting declares that "unless otherwise stated, the auditing standards or practices followed are presumed to be those of the country indicated by the auditor's address" (International Auditing Practices Committe of the International Federation of Accountants 1983, par. 7).

RESULTS OF THE RESEARCH

The study of the accounting principle disclosures in the financial statements of the companies studied revealed that 27 companies used Canadian accounting principles, while four companies used U.S. generally accepted accounting principles. Two companies did not declare the source of their accounting principles. One of the two companies held primarily cash assets and the other reported in U.S. dollars.

Of the 27 firms that used Canadian accounting principles, 24 either provided reconciliations to U.S. accounting principles or disclosed that no material differences existed. Two of the Canadian firms that adopted U.S. generally accepted principles in preparation of their financial statements voluntarily included reconciliations with Canadian accounting principles. Of note is that the firms not disclosing the accounting principles used nor the reconciliation (if Canadian principles are used) are apparently not in compliance with SEC regulations. A summary schedule of conflicts between U.S. and Canadian generally accepted accounting principles reported in the reconciliations provided by the companies is presented in Table 1. The most common area

Table 1.   List of Reported Canadian Accounting Principles in Conflict with
U.S. Generally Accepted Accounting Principles

Deferral of Revenues and Expenses

General and administrative expenses
Administrative expenses related to mineral properties
Legal expenses during the development stage
Travel expenses during the development stage
Interest and other income
Financial costs related to warrants
Franchise costs
Purchased research and development

Nonrecognition of Gains and Losses

Unrealized exchange gains and losses related to long-term debt
Temporary declines in quoted market value of company's portfolio of investments

Amounts Charged Directly to Deficit Account

Write off of deferred exploration and development expenditures
Costs related to the takeover of the company
Prospectus and share issue costs (instead of reducing credit to capital stock)

Asset Valuation

Carrying value of mineral properties acquired from an affiliate in exchange for
     shares ascribed by the board of directors
Valuation of technology rights and fixed assets
Valuation of shares issued for property acquired
Temporary loss on a portfolio of long-term marketable securities in an
     unclassified balance sheet charged to operations instead in stockholders'
     equity valuation account

Financial Statement Disclosures

Cash restricted from general use included in cash and short-term deposits
Subscriptions receivable shown as an asset
Earnings per share calculations

of conflict is the deferral of revenues and expenses in the development stage
which, under U.S. generally acceptable accounting principles, require
immediate recognition. Several Canadian firms report no net income under
this practice of deferral but would show substantial net losses under U.S.
practice. However, three firms reported lower net income under Canadian
principles. Three others disclose a change in accounting principle from
deferring administrative costs to expensing such costs as incurred. Further, the
firms disclosed a variety of write-offs and expenditures that were charged
directly to the deficit account rather than to the income statement. Certainly,

Table 2. Accounting Standards and Reporting Currency
of Financial Statements

Accounting Principles Base	Number of Statements in Specified Currency	
	Canadian Dollars	U.S. Dollars
Canadian Accounting Principles	22	5
U.S. Accounting Principles	1	3
Not Stated	1	1
Totals	24	9

deferral and direct write-off of start-up costs is an important issue for development stage firms. The other significant issue directly related to new enterprises is the valuation of newly acquired properties in a nonmonetary exchange. Three firms reported such differences, but not always providing information as to why the amounts would differ.

As shown in Table 2, 24 of the 33 firms use the Canadian dollar as the reporting currency. The other nine firms use the U.S. dollar for reporting purposes.

The financial statements of the surveyed companies were accompanied by audit reports under Canadian auditing standards or U.S. auditing standards. A breakdown is shown in Table 3. Indeed, the audit report for 21 companies under Canadian standards were accompanied with special "comments to U.S. readers," 19 of which were signed in Vancouver, British Columbia. In this respect, two firms disclosed significant uncertainties in the footnotes but did not include the special report to U.S. readers. Another firm acknowledged in a footnote that a U.S. audit report would require an explanatory paragraph but also did not include the supplementary report nor the explanatory paragraph. As expected, all of the Canada-type audit reports were issued in Canada. However, one U.S-type audit report was issued in Montreal whereas the other five audit reports under U.S. auditing standards were issued in the United States.

Table 3. Types and Location of Issuance of Audit Reports

Types of Audit Reports	Vancouver B.C.	Other Canadian City	In U.S.
Canadian Audit Report with Comments to U.S. readers	19	2	
Canadian Audit Reports only	5	1	
U.S. Audit Reports		1	5
Totals	24	4	5

ANALYSIS OF THE FINDINGS

In reviewing the results of the study, alternative reporting standards are analyzed using either a Canadian base or a U.S. base for application of accounting and auditing principles. In this case, the Canadian base is the home country perspective, a parent company view (Table 4). It excludes those situations where the distinction is not determinable. It shows that the use of the home country base for selection of accounting and auditing standards is the most popular.

A comparison of the reporting concepts utilized shows the consistency in the selection of the accounting and auditing bases. As shown in Table 5, 21 firms prepared financial statements in Canadian dollars, using Canadian accounting principles accompanied by audit reports issued in Canada in accordance with Canadian auditing standards. Three firms issued financial statements under U.S. accounting principles accompanied with an audit report issued in the United States in accordance with U.S. auditing standards. Thus, 24 of the 33 firms were consistent in the selection of accounting and auditing bases. Another six firms were consistent in the selection of their accounting and auditing bases but reported in a currency of the other country. That leaves only three firms with significant inconsistency in their accounting and auditing base selections. One used Canadian dollar reporting under Canadian accounting principles with an audit report in accordance with U.S. auditing standards issued in Canada. The other two did not disclose their accounting principles base. One of the two used U.S. dollar reporting with a Canadian

Table 4. Summary of Reporting Concepts Applied

Reporting Issue	Type of Reporting Base Applied	
	Canadian	United States
Accounting Principles Base	27	4
Reporting Currency Base	24	9
Audit Standards Base	27	6
Audit Place	28	5
Average Frequency of Use	27	6

Table 5. Analysis of Reporting Bases by Firm

Type of Reporting Base	Number of Firms
Pure Canadian Base	21
Candian but with US$ Reporting	5
Canadian but with U.S. Audit Report	1
Pure U.S. Base	3
U.S. with Canadian $ Reporting	1
GAAP Base Not Stated	2

audit report issued in Canada and the other used Canadian dollars reporting with U.S. audit report issued in the United States.

SUMMARY AND CONCLUSIONS

Of the 31 firms disclosing, 27 maintained a Canadian base in the selection of their accounting and auditing standards, and only four firms used a primarily U.S. base. Also significant is that 24 of the sample of 33 companies, or 73% of the firms, were consistent in the use of a single financial reporting base. These Canadian companies disclosed their development stage status in accordance with U.S. accounting standards because of the requirements of the U.S. Securities Exchange Commission. Also, 21 firms that maintained a Canadian base did issue a supplementary "Comments for U.S. readers" report by the auditors in accordance with specific Canadian audit guidelines.

The results of this study reveal some significant variations in financial reporting by Canadian development stage enterprises reporting to the U.S. investment market. If Canadian and U.S. principles are to be harmonized to enhance the goal of reciprocity between the two countries, this area of accounting and financial reporting standards needs attention. Also, the special reporting disclosures for developing stage enterprises might be given attention in the international accounting standard setting process. Finally, if comparability of financial statements across borders is to be enhanced, consistency in the selection of accounting and financial reporting standards is necessary.

APPENDIX

Canadian Development Stage Firms Operating in
the United States Securities Market

Year	Company	Year	Company
1988	Alaska Apollo Gold	1988	Guilt-Free Goodies
1988	Bioject Medical	1988	Gunnar Gold Mining
1987	California Gold Mines	1988	Interstrat Resources
1988	Carolin Mines	1987	London Silver Corporation
1988	Centurion Gold Limited	1988	NETI Technologies
1987	Centurion Minerals	1988	Newfields Minerals
1988	City Resources	1987	New Lintex Minerals
1988	Consolidated NRD Resources	1988	North American Metals
1988	Conversion Industries	1988	O'Hara Resources
1988	Cornucopia Resources	1988	Polestar Exploration
1988	DiaEm Resources	1988	Rich Coast Sulphur
1988	Dumagami Mines	1987	Suncor Communications
1988	Dusty Mac Mines	1988	Technigen
1987	Eastmaque Gold Mines	1988	US Precious Metals
1988	Geodome Resources	1988	ZFAX Image Corporation
1988	Golden Quail		

ACKNOWLEDGMENTS

The authors of this study would like to extend their appreciation for the support received from the Peat Marwick Foundation in the conduct of this research. Thanks are also extended to the staff at the Securities and Exchange Commission Chicago library for their cooperation in obtaining the data for this study.

REFERENCES

Canadian Institute of Chartered Accountants. 1981. *Auditing Guideline, "Canada-United States Reporting Conflict with Respect to Contingencies and Going Concern Considerations."* Toronto: CICA (April).

———. 1988. *Auditing and Related Services Guideline, "Canada-United States Reporting Conflict with Respect to Contingencies and Going Concern Considerations."* Toronto: CICA (December).

Crooch, G.M., and B.E. Collier. 1973. Reporting guidelines for companies in a stage of development. *The CPA Journal* (July): 579-588.

Financial Accounting Standards Board. 1975. *Accounting and Reporting by Development Stage Enterprises, FASB Statement No. 7.* Stamford, CT: FASB. (June).

Goodman, H., and L. Lorensen. 1984. *Illustrations of Accounting and Reporting by Development Stage Enterprises.* New York: American Institute of Certified Public Accountants.

International Auditing Practices Committee of the International Federation of Accountants. 1983. *International Auditing Guideline No. 13: The Auditor's Report on Financial Statements.* New York: IFAC. (October).

Most, K.S. 1984. *International Conflicts of Accounting Standards* (Research Monograph Number 8). Vancouver, B.C.: The Canadian Certified General Accountants' Research Foundation.

Rueschhoff, N.G. 1973. U.S. dollar based financial reporting of Canadian multinational corporations. *The International Journal of Accounting* (Spring): 103-9.

U.S. Securities and Exchange Commission. 1988. *Regulation S-X.* (November).

ACCOUNTING REGULATORY DESIGN:

A NEW ZEALAND PERSPECTIVE

G.D. Tower, M.H.B. Perera , and A.R. Rahman

ABSTRACT

The level of compliance with professional accounting rules and standards is influenced by two important factors: the enforceability of those standards and their acceptability to preparers and stakeholder groups. Several studies have demonstrated that compliance with the professional accounting standards in New Zealand is very low. This study proposes to address this issue. In the paper we examine the process of setting and enforcing accounting standards in New Zealand to: (1) consider the role of the New Zealand Society of Accountants (NZSA) in the regulation of corporate reporting, (2) assess the openness of the standard-setting process, (3) analyze the enforcement procedures, and (4) explore accounting regulatory design alternatives within the New Zealand context. The results of the study indicate that the standard-setting process in New Zealand is not open and lacks adequate representation from groups such as preparers and stakeholders. Moreover, the enforcement mechanism for standards is

Advances in International Accounting,
Volume 5, pages 115-142.
Copyright © 1992 by JAI Press Inc.
All rights of reproduction in any form reserved.
ISBN: 1-55938-415-8

ineffective. It is recommended in this paper that the representation of preparer and stakeholder groups in the standard-setting process be expanded to enhance the level of acceptability of standards, and that the standards be legally backed to improve their enforcement. Enhanced acceptance together with improved enforcement, it is concluded, will lead to greater compliance.

INTRODUCTION

The level of compliance with professional accounting rules and standards is influenced by two important factors: the enforceability of those standards and their acceptability to preparers and stakeholder groups (Masel 1983; Stamp 1979). Accounting rules and standards that are not generally accepted and/ or do not have strong enforcement capabilities are unlikely to generate the type of information sought by their makers due to low levels of compliance. From an accountability point of view, the right of stakeholder groups to be informed about the business activities, and the right to verify the information received, usually via the independent auditor (see Figure 1), are two essential elements of corporate reporting. In other words, factors that cause inadequate information in corporate reports jeopardize attempts at achieving accountability objectives of such reports.

Accounting regulation in New Zealand has three main sources. First, the legal mandate arises from core company law which is considered old and out-of-date (Hickey 1989). The *Companies Act 1955* requires that companies provide a true and fair view of their financial affairs on an annual basis. Second, the government has delegated power over the country's accountants to the New Zealand Society of Accountants (NZSA) which is the only professional body in the country. The *New Zealand Society of Accountants Act of 1958* states that the NZSA is required "[t]o control and regulate the practice of the profession of accountancy in New Zealand" (Section 3 (4) (a)). At present, the NZSA's rules govern only the conduct of members of the profession. In carrying out this responsibility the NZSA promulgates standards and guidelines that cover disclosure and measurement issues. Third, the New Zealand Stock Exchange (NZSE) provides additional disclosure rules for listed companies and the Securities Commission also requires certain disclosures from public issuers.

The current regulatory structure in New Zealand does not appear to have resulted in a high level of compliance with accounting standards by preparers of financial statements. Various studies contained in Ryan (1990a) have documented a historical trend of noncompliance at a high level. For example, Tower, Rahman, Tan, and Cuthbertson (1990) found the noncompliance rate to be 32-48% for the depreciation and subsequent events standards.

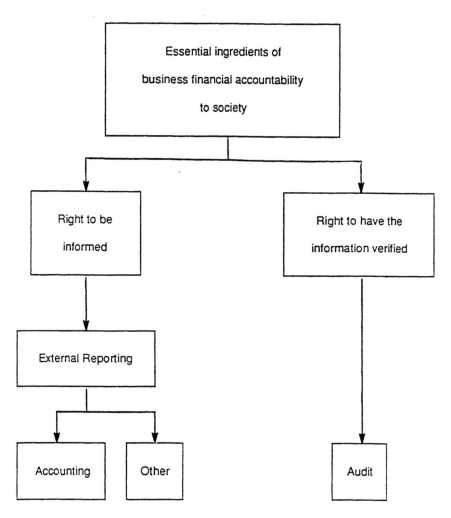

Figure 1. Accountability Framework of Corporate Reporting

The paper considers accounting regulation as an instrument of accountability and, therefore, is not concerned with the general arguments for and against the need for regulation in accounting. It is beyond the scope of this paper to address the issues pertaining to either the need for convincing arguments explaining why a particular standard should be adopted instead of merely prescribing certain accounting methods without any justification, or the importance of the language used in drafting the standards to avoid ambiguities, although these are significant factors contributing to the success or otherwise of the accounting regulatory mechanism as a whole.

This paper proposes to examine the issue of company compliance with accounting rules. In the next section the Anglo-American accounting regulatory structures are analyzed in terms of their strengths and weaknesses. We then offer a discussion of the impact of professional self-regulation on standard-setting structures. In the following section the New Zealand accounting rule-making body is critiqued. This will be done by (a) considering the role of the NZSA in the regulation of corporate reporting, (b) reviewing the NZSA's rule-making process, its openness and representation which, it is suggested, affects the level of acceptability of accounting standards, and (c) analyzing the types of enforcement procedures set in place. A discussion of recent proposals for future direction is also given. An exploration of accounting regulatory design alternatives within the New Zealand context is then offered, followed by the summary and conclusions.

ANGLO-AMERICAN STANDARD SETTING STRUCTURES

The evolution of the New Zealand accounting environment is a result of dynamic interchanges between global, regional, domestic, and internal influences on the standard-setting process (see Figure 2).

One prominent analytical method in identifying such global and regional interchanges is to classify countries into zones of influence. Two methodologies for national comparisons of accounting practices have evolved: one uses an indirect approach by studying economic variables and environmental factors of countries while the other utilizes a direct study of their accounting practices. Both types of classification system have reached the same conclusions concerning New Zealand accounting practices within the Anglo-American model (see, for instance, Frank 1979; Nair and Frank 1980; Amenkhienan 1986; Nobes 1987). Zeff (1979) and Porter (1991) agreed that New Zealand accounting practices (and regulatory structures) are heavily influenced by the Anglo-American countries (Australia, United Kingdom, Canada, and the United States). Tower and Perera (1989) argued that there is a distinct possibility of a regional accounting cluster emerging as a result of the Closer Economic Relations Agreement between Australia and New Zealand.

Each of the Anglo-American countries has attempted to ensure the availability of an adequate level of business enterprise information to the public; they all use regulation of some type to encourage/ensure the level deemed necessary. Yet, one of the most striking features is the wide range of regulatory structures exhibited by these bodies (see Peirson 1990a, 29). This of course is to be expected as accounting is a product of the environment in which it operates (Perera 1989).

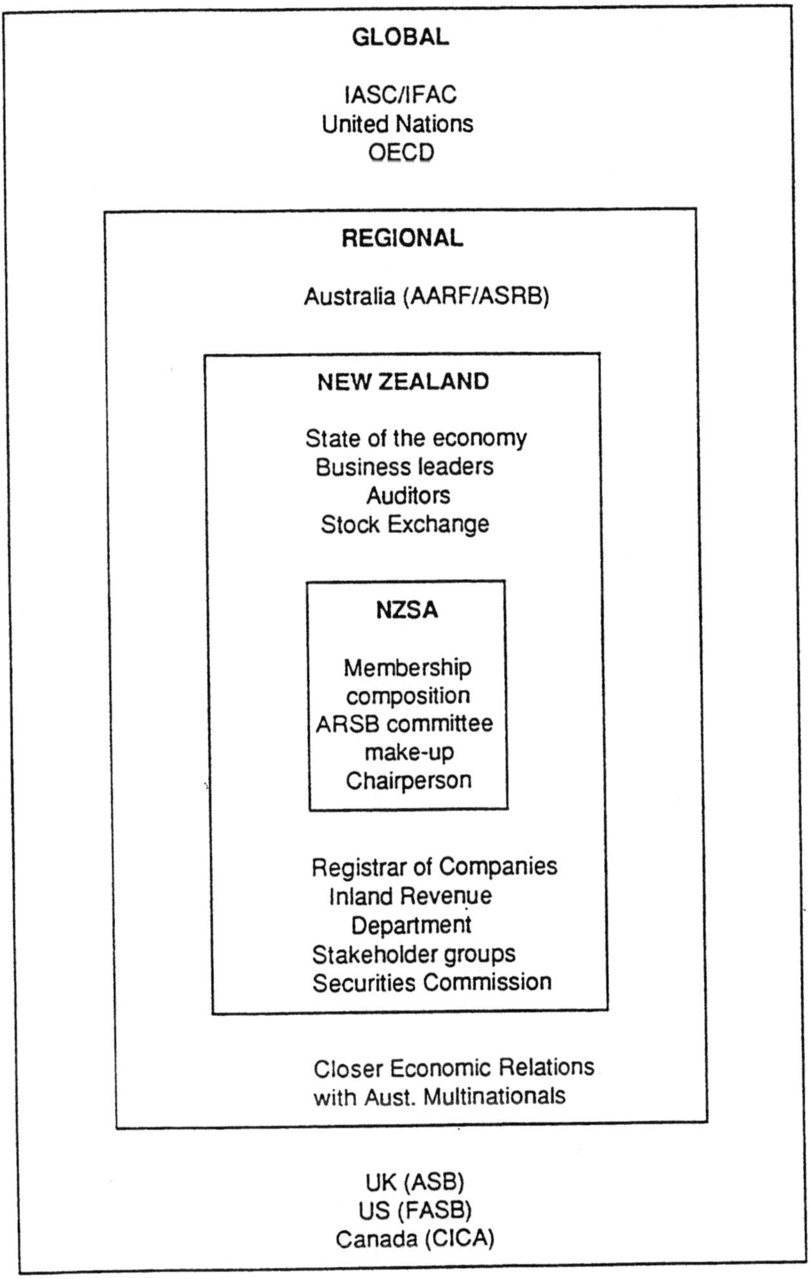

GLOBAL

IASC/IFAC
United Nations
OECD

REGIONAL

Australia (AARF/ASRB)

NEW ZEALAND

State of the economy
Business leaders
Auditors
Stock Exchange

NZSA

Membership
composition
ARSB committee
make-up
Chairperson

Registrar of Companies
Inland Revenue
Department
Stakeholder groups
Securities Commission

Closer Economic Relations
with Aust. Multinationals

UK (ASB)
US (FASB)
Canada (CICA)

Figure 2. New Zealand Accounting Standards Within a Global Context

Membership of standard-setting boards is generally part-time, except for the Financial Accounting Standards Board (FASB). New Zealand and Canadian board members do not receive formal remuneration, whereas the FASB members receive substantial remunerations. Funding arrangements range from broadly based (United States) to professional only (New Zealand and Canada). The U.S. and U.K. bodies claim to be independent of the profession. New Zealand maintains the greatest level of professional self-regulation. Canadian rules have force of law status as do Australian approved standards and the American promulgations are backed by the powerful Securities and Exchange Commission, whereas New Zealand standards have no such legal backing.

An analysis of the Anglo-American accounting regulatory structures reveals two important issues; first, the degree of openness and the extent of representation within the rule-making bodies; and second, the degree of enforcement of the promulgations.

Openness of the Standard-Setting Process

There are different levels of openness within the Anglo-American structures. Only the FASB has an active discussion memorandum and public hearings stage. None of the other standard setters is as open in nature. The professional bodies in the United Kingdom, New Zealand, and Canada choose the board members who are almost inevitably professional accountants. In the United States, the Financial Accounting Foundation (FAF), the parent body of the FASB, selects the FASB members. The FAF is composed of both accountants and nonaccountants. In Australia from 1984-1990, the Accounting Standards Review Board (ASRB) members were chosen by the government through the Ministerial Council which is composed of the attorney generals of the six states and the federal attorney general of Australia.[1]

Canada, the United Kingdom, and the United States have formal arrangements for consultative groups. The latter two have relatively high power sharing arrangements with these groups in that they help appoint the board members. In contrast there is no current formal consultative arrangement in Australia, however, the profession is considering a major overhaul of the standard-setting structure (Peirson 1990a, 1990b) toward a combination of an American and Canadian style system. Part of the change[2] would incorporate some form of Consultative Group. However, the Peirson Report has been criticized recently for the lack of true outside representation in the proposed structure (Starr 1990; Prentice 1990; Walker 1990).

The stated purpose of these consultative groups is to garner feedback from interested parties. It is argued that the meetings of such groups are available only to certain powerful interest parties. For instance, the New Zealand Consultative Group consists of members from the Institute of Directors, Financial Executives Institute, Society of Investment Analysts, Listed

Companies Association, Securities Commission, and the New Zealand Stock Exchange. Other stakeholders of business firms, such as employees, consumer groups, and environmental groups, have no representation and have virtually no impact on the accounting standard-setting process. Furthermore, the consultative group has not held a meeting in the last two-and-a-half years.[3]

Much of the interaction of standard-setting goes on behind closed doors, for instance, Sikka, Willmott, and Lowe (1989) detailed the problems they encountered in attempting to solicit information from the British Accounting Standards Committee (ASC). Typically all decisions within the standard-setting body such as the agenda-setting and formation of options in regards to specific issues are made internally until one late exposure draft phase. For example, in New Zealand there is hardly any outside consultation of many stakeholder groups until Step 7 of a 12-Step process (see Porter 1990, 27). Horngren (1973) emphasised the ramifications of such an approach. He used the development of a marketable securities standard in the United States as a case study of the procedures of the now defunct Accounting Principles Board (APB). He showed how several important changes occurred to the proposed rule before it ever reached the exposure draft phase. Horngren concluded that the options shown to the general public at the exposure draft phase were but a small subset of the original possibilities. This example is especially relevant due to the structural similarities between the APB to the British ASC, Canadian AcSC, and New Zealand ARSB.

The NZSA's mechanism for generating accounting standards is still essentially a "closed shop" operation. The proposed standards are created by the Accounting Research and Standards Board (ARSB), assistance is provided through its committees, especially the Financial Accounting Committee (FAC) and Auditing Committee (AC). Final approval is granted through the Council of the New Zealand Society of Accountants (Porter 1990, 26). No public hearings are held nor are the minutes of meeting or Council votes on issues available for inspection.

Politics has an inherent place within the standard-setting process (Solomons 1978, 1983; Wyatt 1990). Yet with the possible exception of the FASB, none of the Anglo-American standard-setting bodies has a transparent rule-making process in which the interactions and compromises that take place in the process can be viewed. Furthermore, nonaccountants play only a minor role in the entire process.

Enforcement Mechanisms

Canada has the most straightforward enforcement design, with the professional standards being granted force of law status. The FASB's rules are given added weight through the disciplinary powers of the SEC. The United Kingdom recently instituted a new rule-making system with the new *Companies*

Act 1990 containing the power to amend defective accounts. Under this act, court remedies for revision of inadequate accounts and reports can be sought by regulatory authorities. As Walker (1985) pointed out, registration procedures wherein the regulatory institution has the ability to remedy defective accounts clearly give the regulatory agencies much greater authority over companies than do disclosure requirements which are currently in use in Australia and New Zealand. The Australian standards approved by the ASRB also have legislative backing. New Zealand has maintained the profession's predominant role, with no legal backing for NZSA's accounting standards.

There appears to be a serious common problem with regard to the level of compliance with accounting rules. New Zealand-based studies highlight this problem. Eight separate studies examined nine New Zealand accounting standards over a number of years.

Table 1 details an average of 49% noncompliance with New Zealand accounting standards. Walker argued, "[t]he production of accounting rules will be nothing more than symbolic behaviour unless it is accompanied by some programme for monitoring compliance with those standards, and for imposing sanctions for noncompliance" (1985, 12). Similar sentiments have been expressed about the British ASC (ASSC 1975; ICAS 1988; Taylor and Turley 1986).

This high level of noncompliance is even more remarkable given the tremendous latitude in most New Zealand accounting standards. Tower and Bauer (1991) evolved an Ambiguity Index which incorporated the level of defined versus undefined concepts along with permissive versus mandatory phraseology. The evidence generated in that study points to a consistently high degree of ambiguity[4] contained within these rules. This finding is consistent with earlier overseas criticisms offered by Walker (1986) and Thornton (1979).

To diminish this historical trend of high noncompliance the NZSA has two primary enforcement vehicles: censure and the issuance of qualified/ adverse audit opinions. Despite the high level of noncompliance discussed above, the number of qualified or adverse audit opinions is quite low. Fairfield (1990) examined the top 100 New Zealand companies for the periods 1987-1989 and found only an average of 8% qualifications of which approximately three-fourths related to departures from accounting standards. The society has never expelled a member for failure to comply with accounting standards (Zeff 1988).

The apparent lack of enforcement of accounting standards in New Zealand highlights the weakness of professional self-regulation. The NZSA does have a Professional Standard Committee, however, and they have openly decided that the Committee's role is educational rather than disciplinarian (see, for example Edwards 1989).

Table 1. Compliance Studies: New Zealand Company Financial Reporting Practices

Authors	Year Published	Study Period	SSAP Studied	Topic	Degree of Compliance	Overall Averages	Breakdown Time Period	
							1975-1984	1985-1989
Sattler	1978	1975	1	Accounting Policies	40			
Grant et al.	1976	1976	1	Accounting Policies	100			
Sattler	1978	1976	1	Accounting Policies	100			
Sattler	1978	1977	1	Accounting Policies	100			
Ryan	1990b	1987	1	Accounting Policies	63			
Ryan	1990b	1988	1	Accounting Policies	70			
Ryan	1990b	1989	1	Accounting Policies	62	SSAP 1 Average	76.4	85.0	65.0
Grant et al.	1976	1976	3	Depreciation	61			
Tower et al.	1990a	1987	3	Depreciation	69			
Tower et al.	1990a	1988	3	Depreciation	64			
Tower et al.	1990a	1989	3	Depreciation	59	SSAP 3 Average	63.2	61.0	64.0
Grant et al.	1976	1976	4	Inventories	59	SSAP 4 Average	59.0	59.0	
Tower et al.	1990	1987	5	Subsequent Events	52			
Tower et al.	1990	1988	5	Subsequent Events	59			
Tower et al.	1990	1989	5	Subsequent Events	61	SSAP 5 Average	57.3	57.3	57.3
Carslaw and Neale	1990	1989	8	Business Combinations	62	SSAP 8 Average	62.0	62.0	62.0
Cliffe et al.	1984	1981	12	Income Tax	36			
Cliffe et al.	1984	1982	12	Income Tax	48			
Cliffe et al.	1984	1983	12	Income Tax	52	SSAP 12 Average	45.3	45.3	
Emery	1990	1987	13	R&D	73			
Emery	1990	1988	13	R&D	64			
Emery	1990	1989	13	R&D	71	SSAP 13 Average	69.3		69.3
Cliffe	1990	1987	18	Leases	11			
Cliffe	1990	1988	18	Leases	12			
Cliffe	1990	1989	18	Leases	26	SSAP 18 Average	16.3	16.3	16.3
Peterson et al.	1984	1983	CCA1	Inflation Accounting	8	CCA 1 Average	8.5	8.5	
Cameron & Hazelwood	1985	1984	CCA1	Inflation Accounting	9	Compliance	50.8	51.8	55.6

PROFESSIONAL SELF-REGULATION

Professional self-regulation, if successful, could be seen to achieve the goals of accountability without the need for external action. Hines (1989) detailed the history of accountants obtaining professional status as a group of individuals that displayed specialist skills and exhibited a willingness to impose a higher degree of regulation and ethical behavior on themselves. There is disagreement as to whether the primary motivation of professions is to serve the public interest or to enhance their own personal wealth (Mitnick 1980). For example, Wilmott (1986) describes that the primary purpose of professional associations in the United Kingdom is to define, organize, secure, and advance the interests of their own members.

The history of the NZSA as a professional body may be seen in two different ways. On the one hand, many examples can be found of selfless volunteer work serving as technical consultants on such issues as taxation reform and company legislation. Yet, the NZSA could also be seen as a group that protects its own interest. For example, Wilson (1983, 226) argued that accounting standards are the New Zealand profession's security blanket in providing protection against legal action. In 1933, the society was accused by some members of Parliament of running a "closed shop" (Graham 1960, 70). In the formulation of *Companies Act 1955*, the NZSA fought the English Society's desire to allow Commonwealth (i.e., non-New Zealand) accountants to perform audits in New Zealand (Graham 1960, 123). Professional status was reinforced in the *New Zealand Society of Accountants Act of 1958* which gave the NZSA control and regulation over accountants.

The above incidents could be seen as partial evidence for an industry seeking to maximize wealth by restricting entry. This is in essence an industry protection theory as discussed by Stigler (1971), Posner (1974), and Walker (1987) among others. In fact, arguments concerning entry requirements into the accounting profession encouraged the NZSA to formulate long-term objectives and policies on education (NZSA 1984).

The accounting profession's freedom to operate is now at issue. For example, the view that accounting matters are too important to be left for the accountants alone to decide has figured prominently in recent discussions. Further, the threat of deregulation seems to have encouraged the NZSA to at least appear to be more open to the public. Eglinton, the then President of the NZSA, expressed his concern about possible government interference:

> There is a need for us to demonstrate to the legislators and the general public that as a professional body we are capable of operating our own system of quality control. This is the age of consumerism and we must act to put our house in order before someone else tries to do it for us (1988, 5).

However, the lack of public input within the accounting standard-setting process seems in direct conflict with Eglinton's words above. Furthermore, the behind-the-doors approach to standard-setting appears to run counter to the society's own accountability doctrine espoused for public sector accounting (NZSA 1987).

Many of the stakeholder/preparer groups have no representation on the standard-setting structure. There has been *no* nonaccountant representation on the NZSA's ARSB, FAC, AC, or council in at least the last twenty years.[5] Preparer groups have historically been under-represented while public accountants have consistently had over-representation (see Tables 2A-2E).

Tables 2A-2C show that while public accountants historically comprise about 25% of the NZSA membership, they have well over 50% membership of the boards discussed above except for the FAC. Preparers, on the other hand, are largely under-represented (see Table 2D for a variance analysis). Furthermore, Hope and Gray (1982) highlighted the powerful role played by the chairmen of the standard-setting boards; in New Zealand public accountants have been 77% of the council presidents (1970-1991). They have also garnered *all* of the ARSB, FAC, and AC chair positions (1980-1991). Before 1980, the latter three committees were contained within the Board of Research (BOR). The BOR chairpersons for the time period 1970-1979 were also 100% public accountants (see Table 2E).

The maintenance of standards of performance is a central responsibility of any profession (McLean 1984, 82). The Current Cost Accounting (CCA) debacle demonstrated the danger of promulgating rules that are not accepted. Corporate compliance with the CCA standard was only 8-9% and was withdrawn due to the high level of noncompliance (Peterson, Gan and Lim 1984; Cameron and Heazlewood 1985).

There is a general trend of governments becoming more actively concerned with examining the accounting standard-setting process (Flint 1980). A likely scenario for increased New Zealand government involvement would follow a series of major company crashes accompanied by growing public disquiet, public awareness of white-collar crime, press sensationalism, and a clamor for politicians to do something. Most of these factors are already present in the New Zealand environment (see Tables 3 and 4).

Tables 3 and 4 highlight the growing concern caused by the 1987 stock market crash, the seemingly endless list of major companies which were going into receivership or liquidation coupled with collapses such as the Development Finance Corporation (DFC), and huge write-offs by the Bank of New Zealand. The government responded to these pressures by instigating a number of Ministerial Reviews charged with the responsibility for investigating various aspects of corporate behavior, including external financial reporting.

Table 2A. NZSA Accounting Standard Setting Membership Breakdown

	1991	1990	1989	1988	1987	1986	1985	1984	1983	1982	1981	1980	1979	1978	1977	1976	1975	1974	1973	1972	1971	1970
Council																						
Academics	2	1	2	2	2	1	1	1	1	1	1	1	1	1	1	0	0	0	0	0	0	0
Public Acct	15	17	18	15	16	18	18	16	16	16	16	16	16	16	16	17	16	16	16	15	15	17
Company Acct	5	4	2	4	5	3	2	3	4	3	3	3	3	3	3	3	3	3	3	4	4	4
Government Acct	1	1	1	1	–	–	–	–	–	–	–	–	–	–	–	–	–	–	–	–	–	–
Other Acct	–	–	–	–	0	0	0	0	0	0	0	0	0	0	0	0	0	0	0	0	0	0
Total	24	24	24	23	24	23	22	21	22	20	20	20	20	20	20	21	20	20	20	20	20	22
Accounting Research and Standards Board																						
Academics	6	6	6	–	–	–	2	2	2	2	2	3	5	6	6	6	5	5	5	5	5	5
Public Acct	6	6	6	8	8	8	7	7	6	7	7	6	5	5	4	4	5	5	6	5	4	3
Company Acct	2	2	2	–	1	2	2	2	3	3	3	3	2	1	2	2	2	2	1	1	2	2
Government Acct	2	2	–	2	–	–	–	–	–	–	–	–	–	–	–	–	–	–	–	–	0	0
Other Acct	–	1	2	2	0	0	0	0	0	0	0	0	0	0	0	0	0	0	0	0	0	0
Total	12	12	12	12	11	12	12	12	12	12	12	12	12	12	12	12	12	12	12	11	11	10
Financial Accounting Committee																						
Academics	2	2	2	2	2	1	2	2	2	2	2	2										
Public Acct	3	3	3	2	3	3	2	4	4	4	5	4										
Company Acct	3	3	3	4	2	2	0	0	1	1	0	0										
Government Acct	0	0	0	0	0	0	0	0	0	0	0	0										
Other Acct	0	0	0	0	2	0	2	0	0	0	0	0										
Total	8	8	8	8	9	6	6	6	7	7	7	6										
Auditing Committee																						
Academics	2	2	2	2	2	1	1	1	0	0	0	0										
Public Acct.	6	5	5	5	4	4	5	3	6	6	5	5										
Company Acct.	0	0	0	0	0	0	0	0	0	0	0	0										
Government Acct.	1	1	1	1	1	1	0	2	1	1	1	1										
Other Acct.	0	0	0	0	2	0	0	0	0	0	0	0										
Total	9	8	8	7	6	6	6	6	7	7	7	6										
Chairman																						
Council	P	G	P	P	P	C	P	P	P	P	P	P	C	C	P	P	G	P	P	P	C	P
ARSB	P	P	P	P	P	P	P	P	P	P	P	P	P									
Fin. Acct. Cm.	P	P	P	P	P	P	P	P	P	P	P	P		P	P	P		P		P	P	P
Audit Comm.	P	P	P	P	P	P	P	P	P	P	P	P										
BOR	P	P	P	P	P	P	P	P	P	P	P	P										P

Note: P = Public Accountant; G = Government Accountant; C = Company Accountant.

Table 2B. Percentage Occupational Breakdown of NZSA Membership, 1980-1990

	1991	1990	1989	1988	1987	1986	1985	1984	1983	1982	1981	1980	1979	1978	1977	1976	1975	1974	1973	1972	1971	1970
Council						.19	.18	.19	.20	.20	.21	.22	.22	.23	.23	.24	.25	.24	.25	.25	.25	.25
Academics	.01	.01	.01	.02	.02																	
Public Acct	.27	.27	.27	.27	.27																	
Company Acct	.30	.31	.31	.32	.32																	
Government Acct	.07	.07	.07	.07	.07																	
Other	.35	.34	.34	.32	.32																	
Total	1.00	1.00	1.00	1.00	1.00																	

Note: Before 1986, the NZSA only provided information on public accountants *v.* nonpublic accountants.

Source: New Zealand Society of Accountants' annual reports (1970-1990).

Table 2C. Breakdown of Occupational Representation on NZSA Standard Setting Boards
(In Percent)

	1991	1990	1989	1988	1987	1986	1985	1984	1983	1982	1981	1980	1979	1978	1977	1976	1975	1974	1973	1972	1971	1970
Public Accountants																						
PA on Council	.63	.71	.75	.65	.67	.78	.62	.76	.73	.80	.80	.80	.80	.80	.80	.81	.80	.80	.80	.75	.75	.77
PA on ARSB	.50	.50	.50	.67	.73	.67	.58	.58	.50	.58	.58	.50										
PA on FAC	.38	.38	.38	.25	.33	.50	.33	.67	.57	.57	.71	.67										
PA on AC	.67	.63	.63	.71	.67	.67	.83	.50	.86	.86	.83	.83										
PA on BOR													.42	.42	.33	.33	.42	.42	.50	.45	.36	.30
Company Accountants																						
CA on Council	.21	.17	.08	.17	.21	.13	.09	.14	.18	.15	.15	.15	.15	.15	.15	.14	.15	.15	.15	.20	.20	.18
CA on ARSB	.17	.17	.17	.08	.09	.17	.17	.17	.25	.25	.25	.25										
CA on FAC	.38	.38	.38	.50	.22	.33	.00	.00	.14	.14	.00	.00										
CA on AC	.00	.00	.00	.00	.00	.00	.00	.00	.00	.00	.00	.00										
CA on BOR													.17	.08	.17	.17	.17	.17	.17	.09	.18	.20

Note: PA = Public Accountants; CA = Company Accountants; ARSB = Accounting Research and Standards Board; FAC = Financial Accounting Committee; AC = Auditing
Committee.

Source: New Zealand Society of Accountants' annual reports (1970-1990).

Table 2D. Variance Analysis of Council Membership of Public Accountants versus Company Accountants

	1991	1990	1989	1988	1987	1986	1985	1984	1983	1982	1981	1980	1979	1978	1977	1976	1975	1974	1973	1972	1971	1970
% PA on Council	.63	.71	.75	.65	.67	.78	.82	.76	.73	.80	.80	.80	.80	.80	.80	.81	.80	.80	.80	.75	.75	.77
% PA in NZSA	.27	.27	.27	.27	.27	.27	.19	.18	.19	.20	.20	.21	.22	.22	.23	.23	.24	.25	.24	.25	.25	.25
Variance	.36	.44	.48	.38	.40	.51	.63	.58	.54	.60	.60	.59	.58	.58	.57	.58	.56	.55	.56	.50	.50	.52
% CA on Council	.21	.17	.08	.17	.21	.13																
% CA in NZSA	.30	.30	.31	.31	.32	.32																
Variance	-.09	-.13	-.23	-.14	-.11	-.19																

Note: 1991 occupational breakdown percentages are estimates.
      PA = Public Accountants; CA = Company Accountants.
Source: New Zealand Society of Accountants' annual reports (1970-1990).

Table 2E. NZSA Breakdown of Chairman Positions by Occupations
(In Percent)

	Time Period	Public Accountant	Company Accountant	Government Accountant	Academic	Nonaccountants	Total
President-Council	1970-1991	0.77	0.14	0.09	0	0	1.00
Chairmen-ARSB	1980-1991	1.00	0	0	0	0	1.00
Chairmen-FAC	1980-1991	1.00	0	0	0	0	1.00
Chairmen-AC	1980-1991	1.00	0	0	0	0	1.00
Chairmen-BOR	1970-1979	1.00	0	0	0	0	1.00

Note: ARSB = Accounting Research and Standards Board; FAC = Financial Accounting Committee; AC = Auditing Committee; BOR = Board of Research.

Source: New Zealand Society of Accountants' annual reports (1970-1990)

Table 3. New Zealand Economic Condition

Calendar Years	Unemployment (average for year) (000s)	Bankruptcies	Suicides	Percent of Listed Companies Reporting losses (end-year)	Growth in GDP
1980	40	808	337	NA	1.2
1981	45	557	320	NA	3.3
1982	50	569	364	NA	1.9
1983	75	889	352	7.0	1.6
1984	65	814	389	8.5	8.7
1985	61	869	338	8.9	1.6
1986	65	965	414	12.0	1.9
1987	88	1229	463	17.5	0.9
1988	118	1874	NA	35.9	0.7
1989	151	1921	NA	47.6	NA

Source: Catt (1990, 25)

Table 4. Effect of 1987 Share Market Crash in New Zealand

"Substantial loss of investor confidence" (NZSE 1988, 1)
"Value of shares traded fell from $4.46 billion to $1.7 billion" (NZSE 1988, 1)
"Equity funds raised fell from $2,900 million to $715 million" (NZSE 1988, 1)
"The number of companies on the Official List fell sharply" (NZSE 1988, 1)
"Continued efforts were made to improve surveillance and bring disclosure requirements
  to internationally-acceptable levels" (NZSE 1988, 1)
"Market capitalisation down from $24.2 to $21.4 billion" (NZSE 1988, 7)
"Net loss of 49 New Zealand companies; 51 removed and 2 added (NZSE 1988, 7)
"In the first six months of 1988, 380 companies became insolvent compared with 330
  in the whole of 1987" (World Accounting Report 1988, 17)

THE FUTURE OF ACCOUNTING REGULATION IN NEW ZEALAND

The future direction of accounting regulation in New Zealand is a widely discussed issue at the present time among various parties interested in accounting and financial reporting practices of business enterprises. Three completed Ministerial Reviews examined different aspects of the reporting of financial accounting information. The Russell Committee (RC) studied causes of the share market crash and offered some recommendations; the New Zealand Law Commission (NZLC) offered an update to company law; and the Securities Commission (SC) prepared a brief on financial reporting. Officially, these reviews are designed, in part, to uplift investor confidence and improve financial reporting practices. A more cynical view would describe the proliferation of reviews as a method of the government being seen to be doing something—that is, positive action undertaken to promote the public interest.

The fall of the market value of shares by two-thirds in 1988 (NZSE 1988, 1) provides evidence of the extent to which the public had lost confidence in the capital market. The quality and reliability of accounting information is also being questioned ("Accounting Methods Questioned After Crash" 1988). For instance, MacDonald, the Executive Director of the NZSA, admitted there was an accountant's role in the share market crash (Oudtshoorn 1989, 21). The New Zealand share market fell 36% in October 1987. while prices in other markets gradually recovered, New Zealand prices continued to decline. For instance, Brierley Investments, the then largest New Zealand company share price fell from an 1987 price of $5.20 to a 1991 figure of little over $1. Considering its total issue of over one billion shares, this amounts to a wipe-off of over $4 billion from the market in respect of that company. The New Zealand Barclays Index has fallen from over 3600 in October 1987 to around 1400 in May 1991. It is the post crash decline which most notably distinguished New Zealand from the rest of the world (Tripp 1989, 2). This failure to recover and continuing experience of company failures and problems led to the Russell Committee report (*Report of Ministerial Committee of Inquiry Into the Sharemarket* 1989a).

Russell Committee Proposals

The first report (1989a) resulted from the Ministerial Review, while the second report (*Report of the Sharemarket Inquiry Establishment Unit* 1989b) outlined ways to implement the initial recommendations. The latter report recommended a revamped New Zealand Securities Commission (NZSC) be installed as overseer to the equity markets. This is important in view of the absence of such a function at the present time. Listing requirements would be given the force of law and funding would be increased.[6] The reports advocated sanctions for noncompliance with standards. Under their proposal (1989b, 140) certain regulatory agencies or any security holder or creditor of that issuer could appeal to the statutory authority for remedy of a qualified set of financial statements.[7]

The Russell Committee criticized the quality of corporate reporting in New Zealand:

> One of the major disclosure deficiencies identified by the Committee was the level of noncompliance with New Zealand Society of Accountants SSAPs. Despite a requirement in the listing requirements for companies to comply with SSAPs, it is apparent to the Committee that these standards are not adhered to by all listed companies (1989a, 57).

Two important accounting changes were proposed; first, the Eighth Schedule to the Companies Act be updated; and second, legal backing be awarded to accounting standards which are "approved" by an Accounting Standards

Board (ASB).[8] This new board would resemble the ASRB in Australia by acting solely as a reviewer of standards; the NZSA would retain the primary role of researching and drafting of standards.

A New Companies Act

New Zealand's *Companies Act of 1955*, is also currently under review with a new draft, 1990 Companies Act, offered by the NZLC (1987, 1989, 1990). It appears that New Zealand will adopt a bifurcated regulatory system. The new draft act is basic legislation applying to all companies. Nowhere does the current act prescribe the accounting practices to be used in arriving at the figures that appear in the published financial statements, whereas public issuers will have to comply with legally enforceable accounting rules.

The Companies Bill (1990), currently at New Zealand Parliament Select Committee, is silent as to specific accounting rules therein implicitly maintaining the profession's monopoly control. The overriding provision of true and fair view would still prevail, with its inherent ambiguity continuing to cause confusion as to the role and function of accounting standards. In addition, the draft act essentially leaves enforcement in the hands of shareholder suits, a concept ill suited to consumer protection despite the opinions of respondents to the NZLC 1987 report.[9] The draft act does simplify and clarify a badly out-of-date piece of legislation (Jones 1989, 35; Hickey 1989, 36). However, this proposal pursues a free market approach which is in conflict with consumer protection and the accountability theme of external financial reporting.

Securities Commission Recommendations

The report of the Securities Commission (1990) recommended changes to the status quo, calling for a statutory provision for the revision of defective accounts. In supporting the Russell Committee's position on the subject, it recommended the adoption of an Accounting Standards Board (ASB). Both reports envisioned the ASB would: approve accounting standards, have the power to amend standards, be located outside of the accounting profession's domain, limit the NZSA's representation to a large minority, use accountant's technical expertise, and receive submissions from both the NZSA and others. It may be mentioned here that, as stated above, the draft Companies Act does not currently include a provision for evaluating accounting standards (NZLC 1990). The Securities Commission also properly contends that the ASB monitor qualified audit reports, a provision that would effectively strengthen the role of the auditor as an independent watchdog.

The ministerial reviews discussed above were initiated due to a perceived crisis in the accounting profession and continuing company failures along with the present lack of confidence in the equity markets. The issued reports suggested important changes be made to the regulatory design of the New Zealand accounting standard-setting process.

REGULATORY DESIGN

Regulatory forms exist in the wider environment and are subject to changes and adjustments from time to time in response to the changing conditions and circumstances. For example, Rahman, Perera and Tower (1991) noted the growing importance of the Australia and New Zealand Closer Economic Relations Agreement (CER). An important component of CER is the harmonization of business law; harmonization of accounting standards is therefore crucial (Tower and Perera 1989). Rahman et al. (1991) offered an alternative domestic standard-setting structure with New Zealand joining the ASRB to form an Australasian Accounting Review Board. As an interim step, they suggested that New Zealand join the ASRB with board representation.

The regulatory changes and adjustments are highlighted in Bernstein's (1955) classical work in describing the life cycle of a regulatory commissions. This life cycle is in effect a theory of decay. Bernstein's regulatory bodies, as detailed in Mitnick (1980, 46), had four phases: gestation, youth, maturity, and old age.

As emphasized by Bernstein, changes in regulatory mechanisms are usually caused by a perceived crisis (Wells 1978; Rahman 1992). Donald Kirk, FASB Chairman, described this element:

> The extent of standard-setting will depend on the economic environment-turbulent times will demand more standards than stable times. Without economic stability, government intervention is likely to increase. Instability encourages regulation of economic behavior (quoted in Hepp and McRae 1982, 62).

Davis and Menon's (1987, 196) case study of the Cost Accounting Standards Board (CASB) in the United States provided evidence of the Bernstein model, they felt that the absence of a strongly supportive constituent group was a major factor in the demise of this regulatory body. They argued that,

> Changes in the sociopolitical environment in the late 1970s altered the incentives of elected officials, encouraging them to dismantle agencies or reduce their regulatory ambit.... In the new regulatory environment, an agency that lacks a strong supportive constituency is likely to have to fight hard to retain its funding and survive (1987, 197).

Morley (1985) discussed three phases in accounting standard-setting: the design phase, the approval phase, and the enforcement phase. Plott and Sunder

(1981) pointed out an important paradox in the standard-setting process: the need for knowledgeable people to create rules versus the issue of self-interest and potential for capture of a regulatory process. Broad representation of the stakeholder and preparer groups on the standard-setting body would provide a strong constituent base for its activities and this would help enhance the level of acceptability accorded to the standards by those groups. As Horngren stated when describing the FASB arrangement, "[t]he key to a successful enterprise is to generate a product that is *acceptable* to customers" (1973, 62, emphasis in original).

The design phase of developing standards is important because it effectively sets the agenda for topic selection and therein determines priorities. Other important issues are the call for a system to deal with emerging issues, demand for multi-source rule-creators, need for adequate due process, maintenance of · the technical expertise of the profession, and funding adequacy.

The FASB, CICA, and the new British ASB all have an emerging issues task force to deal promptly with important new issues. The Peirson report in Australia was criticized for ignoring this issue (Walker 1990). New Zealand has no such system, nor does it solicit potential standards from other sources (this is advocated by both the Securities Commission 1990 and the Russell Committee 1989b).

Adequate due process prior to submission for approval is also important. The FASB's policy of holding public meetings, discussion memoranda, and publicized votes would greatly enhance the creditability of the New Zealand system. The design phase is most appropriately handled by those most technically competent (Parker, Peirson, and Ramsay 1987) (i.e., the NZSA). The expertise and willingness of the profession to help cannot be ignored.

Another aspect of accounting standard-setting that needs to be considered is funding. For example, both the Wheat Committee in the United States and the Dearing Committee in the United Kingdom sought to ensure that adequate funding was available for the standard-setting body when they made their recommendations to establish the FASB and ASB respectively. As Zeff stated, without adequate resources, "the quality of the drafting will be made to depend on the uncertain support services likely to be provided by the members' companies or firms" (1988, 20). The NZSA offers a large financial commitment to the standard-setting process (see Figure 3 for a graphical depiction of the rising financial outlay made by the NZSA 1970-1990). In the New Zealand environment, it may be appropriate to expect contributions from the government and preparer/stakeholder groups toward this end.

Under the proposals of the Russell Committee (1989a, 1989b) and the Securities Commission (1990) the approval phase would be separated from the NZSA and given to an independent, government backed body. To achieve the objectives of an accountability framework, this new body should have wide outside nonaccountant representation. This concept differs strongly from most

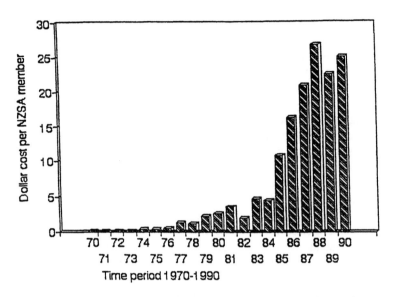

Figure 3.   NZSA Financial Data (1970-1990)

current Anglo-American structures which offer only token outside representation through powerless consultative groups and keep the important decision roles within the profession. However, as Hope and Briggs stated, there is "no obvious reason to assume that accountants with their (necessarily) limited range of expertise, rather than knowledgeable representatives form other groups within society, should control the accounting policy process" (1982, 95).

The proper level of professional representation on a new Accounting Standards Board is problematic. The proposals call for appointments to be made by the Minister of Justice. The Securities Commission called for "some representation for the NZSA" (1990, 67), the NZSA "wished to have the right to nominate some members" (Hays 1990, 1) and the Russell Committee recommended that two of the proposed five board seats should be NZSA nominees (1989b, 150). A limited minority representation which would allow for wider representation from other constituent groups appears appropriate.

As discussed above, the net result of the New Zealand proposals (Russell Committee 1989b; Securities Commission 1990; NZLC 1990), if adopted, will be a split system of strong, widely representative standards for public issuers. Enforcement of accounting rules for the over 150,000 nonpublic issuer companies, however, will continue to be an issue with the lack of any true broad community representation in the standard-setting process. The New Zealand

Companies Act 1955 does not give the NZSA's accounting standards the force of law. The Companies Bill (1990) not only fails to provide this authority but would also remove the disclosure requirements within the Eighth Schedule. In addition, no U.K.-style remedies clause is incorporated nor is the registration method pursued. Shareholder suits would be the supposed main method of enforcement.

As Parker (1986) suggested, the enforcement stage should be given to the entity equipped with the power to enforce, namely the government. Zeff also supports this view when he says that "a government agency is in a much better position to enforce compliance with accounting standards than is the council of a professional accounting body" (1988, 20). The accounting profession does not possess the authority to ensure that the regulation is complied with. The clear logic of this concept underscores the weaknesses of the current NZSA model.

CONCLUSION

This paper set out to provide an up-to-date analysis of the current accounting regulatory design from a New Zealand perspective. The analysis was centered around the issue of compliance with accounting standards, and included discussions on the role of the NZSA and its standard-setting process, the type of enforcement procedures adopted and various accounting regulatory design alternatives. The Anglo-American structures were examined in terms of their strengths and weaknesses with a view to providing a global picture of the issues involved. The impact of professional self-regulation on accounting standards was assessed with a consideration of the alternative regulatory structures available.

The Anglo-American standard-setting bodies were found to be for the most part a "closed-shop" operation with the agenda set by accountants, the rules created by accountants, the rules approved by accountants, applied by accountants, and audited by accountants. Only the United States with the FASB's more visible system and the Australian government's selection of ASRB board members provide significant exceptions. The New Zealand system, even in comparison to the other Anglo-American countries, appears to be particularly ineffective. Professional self-regulation alone, without legal backing and greater community acceptance of their rules, appears inadequate. Various empirical studies have demonstrated the low level of compliance with accounting standards in New Zealand. Wider board representation by stakeholder and preparer groups along with a more transparent process is likely to increase the acceptability of accounting rules by these groups. The increased level of acceptance accompanied with legal backing of accounting standards would lead to a higher degree of compliance.

ACKNOWLEDGMENTS

The authors would like to thank Brenda Porter, participants of accounting seminar presentations at Curtin University of Technology and Massey University, and an anonymous reviewer for their valuable thoughts and comments on earlier drafts of this paper. A grant from the Massey University Research Fund greatly assisted the study.

NOTES

1.   From January 1, 1991 the ASRB has been replaced by the Australian Accounting Standards Board (AASB) (Porter 1991, 78). At the time of writing the composition of the AASB was not known.

2.   Other key points of the Peirson Report include the establishment of an independent foundation and the merger of the ASRB and PSASB (Peirson 1990b).

3.   In a 1991 interview with Kevin Simpkins, the Technical Director for the NZSA, it was discovered that the last meeting of the consultative group was held on June 1, 1988.

4.   In a 1991 interview, Brenda Porter, the former Director of Research for the NZSA, argued that the ambiguity may not be intentional in that the ARSB has a deliberate policy of not issuing narrow, prescriptive standards such as the FASB's rules. Tower and Bauer (1991, 14), however, judged only one contemporary standard unambiguous.

5.   This can be linked back to institutional barriers of NZSA membership criteria, created by the government, in the *New Zealand Society of Accountants Act of 1958.*

6.   They proposed an increase in funding from the current $1.9M to $3.9M.

7.   It is noted that these recommendations are similar to the *UK Financial Services Act 1986.*

8.   The Russell Committee actually titled the proposed board the Accounting Standards Review Board (ASRB) while the Securities Commission proposed an Accounting Standards Board (ASB). To avoid excessive use of acronyms the latter title will be used in this study.

9.   See Appendix 3 of Tower and Perera (1989, 35-36) for the detailed response to this 1987 report.

REFERENCES

Accounting methods questioned after crash. 1988. *The Dominion* (May 3), p. 25.

Accounting Standards Steering Committee. 1975. *The Corporate Report* London: ASSC.

Amenkhienan, F.E. 1986. *Accounting in Developing Countries: A Framework for Standard Setting.* Ann Arbor MI: UMI Research Press.

Bernstein, M.H. 1955. *Regulating Business by Independent Commission.* Princeton, NJ: Princeton University Press.

Cameron, A.F., and C.T. Heazlewood. 1985. Current cost accounting in New Zealand: An analysis of the response to CCA-1. *Discussion Paper No. 33,* Department of Accountancy, Massey University, Palmerston North, New Zealand.

Carslaw, C.A.P.N., and A.Y. Neale. 1990. SSAP-8: Accounting for business combinations. Pp. 104-137 in *New Zealand Company Financial Reporting: 1989,* 4th ed., edited by J.B. Ryan. Levin, NZ: Kerslake, Billens and Humphrey.

Catt, A. 1990. Worse times ahead under reserve bank's regime. *The Dominion* (April 22), pp. 25-27.

Cliffe, C.E. 1990. SSAP 18: Accounting for leases and hire purchase contracts. Pp. 228-240 in *New Zealand Company Financial Reporting: 1989* 4th ed., edited by J.B. Ryan. Levin, NZ: Kerslake, Billens and Humphrey.

Cliffe, C.E., F. Devonport, and A. Robb. 1984. A critical examination of achievements in setting certain accounting standards. *Accountants' Journal* 63(5): 223-225.

Companies Act 1955. 1955. Wellington, NZ: Government Printing House.

Companies Act 1990. 1990. London, UK: HMSO.

Davis, S.W., and K. Menon. 1987. The formation and termination of the cost accounting standards board: Legislative intervention in accounting standard-setting. *Journal of Accounting and Public Policy* 6(3): 185-207.

Edwards, L. 1989. Professional Standard Committee. *Accountants' Journal* 68(1): 10-11.

Eglinton, R. 1988. Presidential report. *New Zealand Society of Accountants Annual Report*. Wellington, NZ: New Zealand Society of Accountants.

Emery, M. 1990. SSAP 13: Accounting for research and development activities. Pp. 209-210 in *New Zealand Company Financial Reporting: 1989*, 4th ed., edited by J.B. Ryan. Levin, NZ: Kerslake, Billens and Humphrey.

Fairfield, A.M. 1990. The audit report. Pp. 267-281 in *New Zealand Company Financial Reporting: 1989*, 4th ed., edited by J.B. Ryan. Levin: Kerslake, Billens and Humphrey.

Flint D. 1980. The significance of the standard 'true and fair view.' *Invitation Research Lecture*, New Zealand Society of Accountants, Wellington, New Zealand (pp. 19-20).

Frank, W.G. 1979. An empirical analysis of international accounting principles. *Journal of Accounting Research* 17(2): 593-605.

Graham, A.W. 1960. *The First Fifty Years, 1909-1959*. Wellington, NZ: New Zealand Society of Accountants.

Grant, A.S., C.S. Greenlees, and B.J. Old. 1976. Public company reporting practices. *Accountants' Journal* 55(11): 400-403.

Hays, P. 1990. Editorial. *Accountants' Journal* (June): 1-3.

Hepp, G.W., and T.W. McRae. 1982. Accounting standards overload: Relief is needed. *Journal of Accountancy* (May): 52-62.

Hickey, E. 1989. Company law reform. *Accountants' Journal* 68(8): 36-40.

Hines, R. 1989. Financial accounting knowledge, conceptual framework projects and the social construction of the accounting profession. *Accounting, Auditing and Accountability Journal* 2(2): 73-92.

Hope, T., and J. Briggs. 1982. Accounting policy making: Some lessons from the deferred taxation debate. *Accounting and Business Research* (Spring): 83-96.

Hope, T., and R. Gray. 1982. Power and policy making: The development of an R and D standard. *Journal of Business Finance and Accounting* (Winter): 531-558.

Horngren, C.T. 1973. The marketing of accounting standards. *Journal of Accountancy* (October): 61-66.

The Institute of Chartered Accountants of Scotland. 1988. *Making Corporate Reports Valuable*. London: Kogan Page.

Jones, D.O. 1989. Company law reform: Observations on the Law Commission's report 'Company law reform and restatement.' *Accountants Journal* 68(6): 35-37.

Masel, L. 1983. The future of accounting and auditing standards. *Australian Accountant* (September): 541-549.

McLean, L. 1984. The setting and enforcement of accounting standards and research. *Accountants' Journal* 63(2): 82-83.

Ministerial Committee of Inquiry into the Sharemarket. 1988. *Call for Submissions*. Wellington, NZ: Department of Justice.

Mitnick, B.M. 1980. *The Political Economy of Regulation: Creating, Designing, and Removing Regulatory Forms*. New York: Columbia University Press.

Morley, B.A. 1985. The ASRB: Its role and current issues. Paper presented at Hobart, Victoria State Congress of the Institute of Chartered Accountants, Australia. (April 18).

Nair, R.D., and W.G. Frank. 1980. The impact of disclosure and measurement practices on international accounting classifications. *Accounting Review* 55(3): 426-450.

New Zealand Law Commission. 1987. Company law: A discussion paper. *Preliminary Paper No. 5.* Wellington, NZ: Government Printing House.

_____. 1989. *Company Law: Reform and Restatement.* Wellington, NZ: Government Printing House.

_____. 1990. *Company Law Reform: Transition and Revision.* Wellington, NZ: Government Printing House.

New Zealand Society of Accountants. 1984. *Horizon 2000–and Beyond: Issues Facing the Accountancy Profession.* Wellington, NZ: NZSA.

_____. 1987. *Statement of Public Sector Accounting Concepts,* (*SPSAC I*). Wellington, NZ: NZSA.

New Zealand Society of Accountants Act of 1958. 1958. Wellington, NZ: Government Printing House.

New Zealand Stock Exchange. 1988. *New Zealand Stock Exchange 1987/1988 Annual Report.* Wellington, NZ: NZSE.

Nobes, C.W. 1987. Classification of financial reporting practices. Pp. 1-22 in *Advances in International Accounting*, Vol. 1, edited by K.S. Most. Greenwich, CT: JAI Press.

Oudtshoorn, N.V. 1989. Profile: Ross MacDonald. *Australian Accountant* (June): 19-25.

Parker, R.H. 1986. Accounting standards and the law: An Australian experiment. *Working Paper No. 18*, University of Sydney Accounting Research Centre, Sydney, Australia (October).

Parker, R.H., C.G. Pierson, and A.L. Ramsay. 1987. Australian accounting standards and the law. *Companies and Securities Law Journal* (November): 231-246.

Peirson, C.G. 1990a. *A Report on Institutional Arrangements for Accounting Standard Setting in Australia.* Melbourne, Australia: Australian Accounting Research Foundation.

_____. 1990b. Reforming standard setting. *Australian Accountant* 50(10): 14-23.

Perera, M.H.B. 1989. Accounting in developing countries: A case for localised uniformity. *British Accounting Review* 21(2): 141-157.

Peterson, R., H.E Gan, and K.L. Lim. 1984. CCA: The day after. *Accountants' Journal* 63(2): 88-97.

Plott, C.R., and S. Sunder. 1981. A synthesis—supplement. *Journal of Accounting Research* 19(Suppl.): 227-239.

Porter, B. 1990. A standard is born. *Accountants' Journal* 69(9): 25-31.

Porter, B. 1991. Update on financial reporting. *Continuing Education Course Paper No. 354*, New Zealand Society of Accountants, Wellington. (May/June).

Posner, R.A. 1974. Theories of economic regulation. *Bell Journal of Economics and Management Science* (Autumn): 335-358.

Prentice, P. 1990. Users may miss out yet again. *Australian Accountant* (November): 24.

Rahman, A.R. 1992. *The Accounting Standard Review Board-The Establishment of Its Participative Review Process.* New York: Garland.

Rahman, A.R., M.H.B. Perera, and G.D. 1991. Tower Toward an accounting regulatory union between New Zealand and Australia. Paper Presentation, Accounting Association of Australia and New Zealand annual conference, Brisbane Australia (July).

Report of Ministerial Committee of Inquiry into the Sharemarket. 1989a. *Sharemarket Inquiry* (known as the Russell Committee). Wellington, NZ: Government Printing Office. (April).

Report of the Sharemarket Inquiry Establishment Unit (known as the Russell Committee). 1989b. Wellington, NZ: Tribunal Division, Department of Justice (November).

Ryan, J.B. 1990a. *New Zealand Company Financial Reporting: 1989*, 4th ed., edited by J.B. Ryan. Levin, NZ: Kerslake, Billens and Humphrey.

_____. 1990b. SSAP 1: Determination and disclosure of accounting policies. Pp. 27-40 in *New Zealand Company Financial Reporting: 1989*, 4th ed., edited by J.B. Ryan. Levin, NZ: Kerslake, Billens and Humphrey.

Sattler, G. 1978. *Statements of Standard Accounting Practice: A Study of The Effect of Their Introduction.* Unpublished paper, School of Management Studies, University of Waikato, Hamilton, New Zealand.

Securities Commission. 1990. *Capital Structure and Financial Reporting in New Zealand.* Wellington, NZ: Government Printing Office.

Sikka, P., H. Willmott, and T. Lowe. 1989. Guardians of knowledge and public interest: Evidence and issues of accountability in the UK accountancy profession. *Accounting, Auditing and Accountability Journal* 2(2): 47-71

Solomons, D. 1978. The politicisation of accounting. *Journal of Accountancy* (November): 65-72.

_____. 1983. The political implications of accounting and accounting standard-setting. *Accounting and Business Research* (Spring): 107-118.

Stamp, E. 1979. The future of accounting and auditing standards. *ICRA Occasional Paper No. 18*, International Centre for Research in Accounting, University of Lancaster, Lancaster, England.

Starr, M. 1990. Independence: An overrated illusion. *Australian Accountant* (November): 26-27.

Stigler, G.J. 1971. The theory of economic regulation. *Bell Journal of Economics and Management Science* 2(1): 3-21.

Taylor, P., and S. Turley. 1986. *The Regulation of Accounting.* Oxford, UK: Basil Blackwell.

Thornton, G.C. 1979. *Legislative Drafting.* Canada: Butterworths.

Tower, G.D., and L. Bauer. 1991. "An empirical study of communication in New Zealand annual reports: Contrasting the public and private sectors. Paper presented at the Accounting Association of Australia and New Zealand annual conference, Brisbane, Australia (July).

Tower, G.D., and M.H.B. Perera. 1989. Closer economic relation (CER) agreement between New Zealand and Australia: A catalyst for a new international accounting force. *Discussion Paper No. 96*, Department of Accounting, Massey University, Palmerston North, New Zealand (September).

Tower, G.D., M. Grosh, A.R. Rahman, L.M. Tan, and J. Cuthbertson. 1990. SSAP-3: Depreciation of fixed assets and measurement of fixed asset. Pp. 44-73 in *New Zealand Company Financial Reporting: 1989*, 4th ed., edited by J.B. Ryan. Levin, NZ: Kerslake, Billens and Humphrey.

Tower, G.D., M. Grosh, L.M. Tan, A.R. Rahman, and J. Cuthbertson. 1990. SSAP-5: Accounting for events subsequent to balance date. Pp. 80-84 in *New Zealand Company Financial Reporting: 1989* 4th ed., edited by J.B. Ryan. Levin, NZ: Kerslake, Billens and Humphrey.

Tower, G.D., A.R. Rahman, L.M. Tan, and J. Cuthbertson. 1990. Most companies fail to explain deviations. *National Business Review* (April 4), p. 13.

Tripp, S. 1989. Accounting standards: A prognosis and prescription. Unpublished masters project, Department of Accountancy, Massey University, Palmerston North, New Zealand.

Walker, R.G. 1985. The ASRB: Policy formation, political activity and research. Paper Presented at Accounting Association of Australia and New Zealand annual conference, Sydney, Australia (August).

_____. 1986. Accounting standards text and context. Paper Presented at the South East Asian University Accounting Teachers Conference, Singapore (April 28-30).

_____. 1987. Australia's ASRB: A case study of political activity and regulatory 'capture.' *Journal of Accounting and Business Research* (Summer): 269-286.

_____. 1990. Reforms: Some fundamental flaws. *Australian Accountant* (November): 16-19.

Wells, M.C. 1978. Is it futile to impose accounting standards. *ASA Research Lecture*, University of Melbourne, Australia (October 26).

Willmott, H. 1986. Organising the profession: A theoretical and historical examination of the development of the major accounting bodies in the UK. *Accounting Organization and Society* 11(6): 555-580.

Wilson, R. 1983. Accounting standards: A legal view. *Accountants' Journal* 62(5): 225-226.

Wyatt, A. 1990. Accounting standards: Conceptual or political? *Accounting Horizons* 4(3): 83-88.

Zeff, S.A. 1979. *Forging Accounting Principles in New Zealand.* Wellington, NZ: Victoria University Press.

————. 1988. January. Setting accounting standards: Some lessons from the US experience. *The Accountant's Magazine*: 20-22.

THE EFFECT OF DISCLOSURE
OF FOREIGN TRANSACTIONS
ON THE EVALUATION OF A FIRM:
EVIDENCE FROM NIGERIA

Daniel P.S. Asechemie

ABSTRACT

Two field experiments were conducted to determine if a Statement of Transactions in Foreign Currency as recommended in the *Corporate Report* would lead to evaluations of financial vulnerability, risk, uncertainty, and stability significantly different from evaluations based on financial statements, including a Value Added Statement, required under SAS 2 in Nigeria. Twenty-nine subjects took part in experiment I, 17 in experiment II. Subjects in each experiment were randomly divided into those that had the SAS 2 statements only and those that had the SAS 2 statements and the Statement of Transactions in Foreign Currency. In experiment I, only the differential evaluations by the two groups of respondents with respect to financial stability was significant, while in experiment II the differentials were significant with respect to all evaluation

Advances in International Accounting,
Volume 5, pages 143-160.
ISBN: 1-55938-415-8

dimensions except stability. These results imply that the Statement of Transactions in Foreign Currency may have incremental information content beyond the SAS 2 Value Added Statement and ought to be required in external reporting. The results also indicate that a high proportion of subjects did not pay attention to the foreign sector of the firm being evaluated. The reason for this neglect is likely to be the unfamiliarity with the Value Added Statement and the Statement of Transactions in Foreign Currency. If so the accounting profession should remedy the unfamiliarity through its educational programs.

INTRODUCTION

The Corporate Report (ASSC 1975) represents a major rethinking on the part of the international accounting profession on the matter of corporate financial reporting and accountability. Since the explication and recommendation of the Value Added Statement by that report, that statement has increasingly become a standard component of financial reporting worldwide. Meek and Gray (1988) discuss progress concerning this statement. Yet the Value Added Statement is not the only one explicated and recommended by the report. It is not clear why the other statements in the *Corporate Report* have received little or no attention from the international accounting community.

One of the statements in the recommended but neglected category is the Statement of Foreign Currency Transactions. Foreign currency transactions, from the point of view of an economy, are important not only because they reflect economic dependencies among nations, but also because they introduce additional variability in the performance of firms coming from the volatility of foreign exchange rates and political machinations. A priori, therefore, a Statement of Foreign Currency Transactions has potentially useful information. In writing the *Corporate Report*, the Accounting Standards Steering Committee saw the potential usefulness of this statement. The basis of the Statement of Transactions in Foreign Currency recommended in paragraph 6.30 of the *Corporate Report* is cash. The following information is recommended for disclosure:

1. cash receipts from exports,
2. cash payments, distinguishing between those of a capital and revenue nature,
3. overseas borrowings remitted to or repaid from the country,
4. overseas investments and loans made from or repaid to the country, and
5. overseas dividends and other payments to capital received from or remitted overseas.

In Nigeria, the Statement of Accounting Standard (SAS) No. 2 recommends the inclusion of the Value Added Statement in Corporate Financial Reports. SAS 2 attempts to capture foreign currency transactions in the Value Added Satement by requiring a breakdown of goods and services bought into Nigerian and foreign components. The SAS does not require a similar breakdown of sales to allow the determination of the foreign currency element of value added. The significance of this omission is far from obvious.

The objective of this study, therefore, is to investigate whether a full-fledged statement on foreign currency transactions would have incremental information content over and above the Value Added Statement recommended in SAS 2. The alternate form of the hypotheses being tested is:

$$\mathbf{H_A}: U_x = U_y$$

where

U_x = the mean evaluation on each dimension of performance in the case where subjects received the Statement of Transactions in Foreign Currency

U_y = the mean evaluation on each dimension of performance in the case where subjects did not receive the Statement of Transactions in Foreign Currency

The critical region at size a is defined by

$$|\bar{x} - \bar{y}| > t_{a/2}(n+m-2) \sqrt{\left[\frac{(n-1)s_x^2 + (m-1)s_y^2}{n+m-2} \right] (1|n+1|m)}$$

where

$\bar{x}$ = sample equivalence of U_x
$\bar{y}$ = sample equivalence of Uy
$t_{a/2}$ = tabled t at size $a/2$
n = sample size for group x
m = sample size for group y
s_x^2 = variance of x
s_y^2 = variance of y (see Hogg and Tanis 1983, 348).

This investigation is conducted by means of field experimentation.

The rest of this paper gives more information about the method of study, presents the results, and discusses these results.

METHODS

Subjects

Mature adults took part in this study. One hundred and fifty adults were contacted by the investigator personally and/or through assistants. Which persons were so contacted depended on chance, but contact effort was limited to the southern states of Nigeria in order to enhance the response rate and reduce study costs. Of the 150 contacted, 46 adults returned responses, representing a response rate of about 30%.

About 96% of the respondents undertook financial analysis in their personal or official capacity or have had a training involving financial analysis. More than 53% of them were accountants; another 20% were bankers, economists, or were involved in other branches of the financial sector. Their ages ranged between 24 and 51 years. A total of 29 subjects took part in experiment I where the questionnaires were accompanied by a full set of financial reports. Because of the low level of use of foreign currency transaction data in that experiment, experiment II was undertaken in which questionnaires were accompanied by either the Value Added Statement or the Satement of Foreign Currency Transactions only. The remaining subjects (17) took part in experiment II.

Research Design

To assess the incremental information content of the Statement of Transactions in Foreign Currency over the SAS 2 Value Added Statement, subjects were divided into two groups. One group was given a full set of financial reports of the experimental company without the Statement of Transactions in Foreign Currency (but including the SAS 2 Value Added Statement), while the other group was given the same reports that the first group had plus the Statement of Transactions in Foreign Currency. The evaluations of the two groups of the riskiness, uncertainty, stability, and vulnerability of the experimental firm were then compared in order to observe any differences in evaluation that may exist. This is experiment I.

A full set of financial reports did not direct the attention of subjects to the primary matter of foreign currency transactions, causing a leakage in the comparison of the SAS 2 Value Added Statement and the Statement of Transactions in Foreign Currency. In experiment II, therefore, only one of these statements was attached to the questionnaire of each experimental group. If the Statement of of Foreign Transactions has any incremental information at all, experiment II provided the best possible environment to observe this.

Procedure

The experimental instrument consisted of a questionnaire to which was attached a set of financial statements. Subjects were asked to study the financial statements and to answer the experimental questions based thereon. Several days were allowed for the subjects to complete their task (some took longer than others). Completed questionnaires were collected personally by the investigator or through assistants.

The primary task subjects were asked to perform was the evaluation of a firm with respect to risk, uncertainty, vulnerability, and stability on a 5-point Likert-type scale running from 1 through 5 (see Appendix for a specimen questionnaire that includes a copy of the accompanying financial statements). These dimensions of evaluation were obtained from paragraph 6.28 of the *Corporate Report* which states:

> In that the degree of risk and uncertainty attached to such (overseas) dealings may be significantly different from that attached to U. K. operations, figures relating to transactions in foreign currency may assist users to assess the stability and vulnerability of the reporting entity.

Instrument

The financial report employed in the experiment belonged to a real-life company. This company was chosen because it was one of a few Nigerian firms that prepared the SAS 2 Value Added Statement for the year ending 1988. The report itself is real, except for the Statement of Transactions in Foreign Currency which was constructed from the actual report as follows: The figure for "Export of Goods and Services" was obtained from a breakdown in the Group Profit and Loss Account, while that for accrual "Import of raw materials" was obtained from the Statement of Value Added. The cash figure for "Import of raw materials" was obtained by subtracting from the 1988 "Deposits for Imports and Remittances" the 1987 figure for unremitted dividends, both balance sheet items. This estimate would be a poor one to the extent that some of the imports were paid for successfully, involving no outstanding deposits on balance sheet date. There was no specific information on other foreign currency revenues and expenses, so these were given zero value. Similarly, capital flows were zero, because there was no indication of either new foreign equity or debt investments.

An effort was made to conceal the identity of the company in order to eliminate possible bias from its identity. That effort consisted of keeping out of the experimental instrument any information suggestive of the identity of the company or the nature of its business.

Of the four evaluation dimensions mentioned in the *Corporate Report*, vulnerability and stability are most likely to invoke very different meanings and responses, because of misunderstanding. Indeed, a pilot questionnaire on the meaning of these terms confirmed this claim. A group of four academic members of the Faculty of Management Sciences of a reputable Nigerian university was gathered to discuss the terms and provide consensual definitions. The consensual definitions have been employed in the development of the questionnaire. Risk and uncertainty were agreed by the academics to have common meanings. It was agreed that risk is commonly taken to mean the possibility of loss; while uncertainty commonly means being unsure of or having a lack of knowledge about an event. It is assumed that subjects adopted these common meanings.

The Statement of Transactions in Foreign Currency was the only statement that differentiated the information received by the two groups of respondents in experiment I. The standard set of financial statements consisted of:

1. a Balance Sheet,
2. a Profit and Loss Account,
3. Notes to the Accounts,
4. a Value Added Statement of SAS vintage,
5. Accounting Policies,
6. a Statement of Source and Application of Funds, and
7. a 5-year Financial Summary.

RESULTS

In experiment I, a significant difference was found in the mean evaluation of stability by subjects with and without the Statement of Transactions in Foreign Currency in a two-sided *t*-test of size .05. A significant difference was absent in respect of risk, uncertainty, and vulnerability in that experiment. These results do not change even at size .01. In the same experiment, it was found, as indicated by the figures in Table 1, that subjects who had the Statement of Transactions in Foreign Currency gave higher evaluations of risk and stability and lower evaluations of vulnerability and uncertainty than those who did not have that statement.

The lack of significance with respect to risk, uncertainty, and vulnerability in experiment I may be attributable to the effect of information in other parts of the annual report offsetting information on the foreign sector of the firm's operations. This possibility is confirmed by the results of experiment II in which differential evaluations were significant at size .05 with respect to vulnerability, risk, and uncertainty, except that, surprisingly, the difference with respect to stability was insignificant. In experiment II, evaluations of subjects with the

Table 1.   Experiment I Statistics

Evaluation Dimension	Case 1 (x)*		Case 2 (y)*		t-statistic at size .05
	Mean Evaluation	Variance	Mean Evaluation	Variance	
Vulnerability	2.86	.84	3.07	1.40	.81
Risk	2.79	1.02	2.73	.99	.76
Uncertainty	2.57	.39	2.73	1.33	.81
Stability	2.86	.55	2.33	.36	.54**
Sample Size	14		15		

Notes: * Case 1 refers to the group of subjects with the Statement of Transactions in Foreign Currency, while Case 2 refers to the group that did not have that statement.
  ** Significant at size .05.

Table 2.   Experiment II Statistics

Evaluation Dimension	Case 1 (x)*		Case 2 (y)*		t-statistic at size .05
	Mean Evaluation	Variance	Mean Evaluation	Variance	
Vulnerability	4.70	.21	3.28	.20	.47**
Risk	4.40	.24	3.42	.24	.51**
Uncertainty	4.50	.45	3.57	.24	.63**
Stability	2.10	1.69	2.57	.52	1.15
Sample Size	10		7		

Notes: * Case 1 refers to the group of subjects with the Statement of Transactions in Foreign Currency, while Case 2 refers to the group that did not have that statement.
  ** Significant at size .05.

Statement of Transactions in Foreign Currency were higher for vulnerability, risk, and uncertainty, but lower for stability than those by subjects without that statement (see Table 2).

An analysis of reasons given by subjects in experiment I for their evaluations is instructive. The analysis showed that subjects with the Statement of Transactions in Foreign Currency gave reasons for their assessment of vulnerability that referred to foreign exchange-related data only once in fifteen responses. Assessments of uncertainty referred once to the Statement of Value Added for the considerable dependence on foreign suppliers for inputs and once to the Statement of Transactions in Foreign Currency for the effect of the exchange rate and the frequent, extensive, and aggressive policy operations of government in the foreign exchange sector. There were no references to the foreign sector elements of either the Statement of Value Added or the Statement of Transactions in Foreign Currency in assessments of risk and stability.

Among respondents who had no Statement of Transactions in Foreign Currency, four referred to the foreign sector in the Statement of Added Value in the assessment of financial vulnerability compared to one among those with the Statement of Transactions in Foreign Currency. In the assessment of risk two respondents referred to the foreign sector where one respondent did so among the other group. As in the group with the Statement of Transaction in Foreign Currency, two respondents referred to the foreign sector in the assessment of uncertainty in the group that did not have that statement. In the assessment of financial stability, there was no reference to the foreign sector just as in the other case with the Statement of Transactions in Foreign Currency. Respondents almost always referred to traditional liquidity ratio in their evaluation of financial stability.

DISCUSSION

Although the results of experiment I may appear inconclusive, especially in view of the analysis of the reasons given for the evaluations made, the result of experiment II do suggest strongly that the Statement of Transactions in Foreign Currency may have incremental information beyond that in the Statement of Added Value, giving rise to significant differences in the evaluation of the financial vulnerability, riskiness, and uncertainty of a firm. The implication of this suggestion is that the Statement of Transactions in Foreign Currency may be introduced as a required part of the body of financial statements for external reporting in Nigeria. This suggestion may still be valid after allowing for the additional disclosure requirements in SAS 2 in respect to foreign entities to which a Nigerian company is affiliated.

It is disappointing that only a small proportion of respondents in experiment I referred to the foreign currency elements of both the Statement of Value Added and the Statement of Transactions in Foreign Currency. In Nigeria, the government is very active in the foreign exchange market and pursues economic diplomacy as a foreign policy thrust. Besides, a large number of unfamiliar foreign economic operations exist that could influence a myriad of variables related to the performance of a firm. Therefore, it does not seem reasonable for financial analysts to ignore the foreign sector of a firm that they are evaluating. One reason why the foreign sector may have been ignored in the present experiments is the relative unfamiliarity of the Statement of Value Added and the Statement of Transactions in Foreign Currency to most of the respondents. In this regard, the accounting profession should popularize these statements among both producers and consumers of accounting information through seminars and continuing professional education programs.

APPENDIX

University of Port Harcourt Faculty of
Management Sciences Department of Accounting

RESEARCH QUESTIONNAIRE

The financial statements accompanying this questionnaire relate to a Nigerian company. They have received a clean audit report from a firm of chartered Accountants. The statements consist of

   i.   a Balance Sheet
   ii.   a Profit and Loss Account
   iii.   Notes to the Accounts
   iv.   a Value Added Statement
   v.   Accounting Policies
   vi.   a Statement of Source and Application of Funds
   vii.   a 5-year Financial Summary
   viii.   a Statement of Transactions in Foreign Currency

Please study the statements

1.   Do you have all the above components in the package of financial statements given to you?   Yes/No
2.   If no, which are missing? _____

_____

3.   How exposed is the company to environmental shocks, that is, to sudden, adverse events? (tick)
   (a)   very highly exposed
   (b)   highly exposed
   (c)   exposed
   (d)   slightly exposed
   (e)   not exposed at all
4.   How much risk is there in the company's operations? (tick)
   (a)   very high risk
   (b)   high risk
   (c)   moderate risk
   (d)   little risk
   (e)   no risk at all
5.   How much uncertainty is there in the company's operation? (tick)
   (a)   very high uncertainty
   (b)   high uncertainty
   (c)   moderate uncertainty
   (d)   little uncertainty
   (e)   no uncertainty at all.

6. What is the capacity (in terms of liquid resources) of the company to meet unforeseen mishaps that could hinder effective performance of its business? (tick)
   (a) very high capacity
   (b) high capacity
   (c) reasonable capacity
   (d) little capacity
   (e) no capacity

7. Why did you respond the way you did to question 3?_____

_____

8. Why did you respond the way you did to question 4?_____

_____

9. Why did you respond the way you did to question 5?_____

_____

10. Why did you respond the way you did to question 6?_____

_____

11. How long did it take respond to questions 3 to 6? _____

_____

12. How mentally and/or computationally exerting is the exercise in responding to questions 3 to 6?_____

_____

13. How old are you? _____ years.
14. What is your profession? _____
15. Do you undertake financial statement analysis in your professional and/or personal capacity? _____

_____

16. Do you have any kind of training in financial statement analysis? __

_____

17. What is your sex?   Male/Female

Group Profit and Loss Account for the year ended September 30, 1988

	Notes	1988 (in 000's)	1987 (in 000's)
GROSS SALES		247,395	1173,566
EXCISE		56,920	39,939
TURNOVER		190,475	133,627
TRADING PROFIT	1	59,215	42,900
INVESTMENT INCOME		716	544
PROFIT			
for the year before taxation		59,931	42,900
TAXATION	2	23,614	15,246

	Notes	1988 (in 000's)	1987 (in 000's)
PROFIT			
for the year after taxation		36,317	27,654
EXTRAORDINARY ITEM	3	92	111
PROFIT			
for Taxation & Extraordinary item		36,409	27,765
TRANSFER TO INFLATION RESERVE		6,750	10,000
PROFIT AVAILABLE FOR DISTRIBUTION		29,659	17,765
DIVIDEND		27,000	13,500
TRANSFER TO OTHER REVENUE RESERVE			
RESERVE	11	2,659	4,265
EARNINGS PER STOCK UNIT	4	36.3k	27.7k

Balance Sheets as at September 1988

		Group		Company	
		1988	1987	1988	1987
ASSETS EMPLOYED	Note	000's	000's	000's	000's
FIXED ASSETS	5	115,952	107,088	115,952	107,088
INVESTMENTS	6	228	228	1,300	1,300
		116,180	107,316	117,252	108,388
CURRENT ASSETS					
Stocks	7	136,826	89,031	136,826	89,031
Debtors		6,080	9,996	6,079	9,993
Deposits (Imports &					
Remittances)		8,062	13,918	8,062	13,918
Bank & Cash Balances		8,091	16,016	8,079	16,008
		159,059	128,961	159,046	128,950
CURRENT LIABILITIES					
Creditors	8	38,021	30,833	38,021	30,832
Amounts due to Group					
Companies	9	20,372	17,186	22,355	19,170
Taxation		29,533	21,668	29,531	21,666
Proposed Dividend		27,000	13,500	24,000	13,500
		114,926	83,187	116,907	85,168
Net Current Assets		44,133	45,774	42,139	43,782
Net Assets		160,313	153,090	159,391	152,170
FUNDS EMPLOYED					
SHARE CAPITAL	10	50,000	25,000	50,000	25,000
RESERVES	11	110,313	128,090	109,391	127,170
Stockholders' funds		160,313	153,090	159,391	152,170

Principal Accounting Policies

(a) *Accounting Convention.* The accounts are prepared under the historical cost convention modified by the revalaution of leasehold properties.

(b) *Basis of Consolidation.* The consolidated accouts incorporate the accounts of the company and its subsidiary.

(c) *Turnover.* Turnover represents the invoice value of sales to externa customers after deducting excise duty.

(d) *Fixed Assets.* Fixed Assets are stated net of depreciation provided thereon to date. Plant and Equipment and Motor Vehicles are stated at cost while Leasehold Properties are stated at cost or valuation. Leases which at the balance sheet date have expected useful lives of less than 50 years are designated short-term leases.

(e) *Depreciation.* The cost of valuation of fixed assets is written off by equal installments over their expected useful lives. The principal annua rates of depreciation are:

Leasehold Properties	2½%
Plant and Equipment	7%
Motor Vehicles	25%

(f) *Investment Income.* The profits of companies in which the group holds shares as trade investments are included only to the extent o dividends received.

(g) *Stocks.* Stocks are stated at the lower of average cost and ne realizable value. In the case of leaf stocks and locally manufactured goods, including a relevant proportion of overheads.

(h) *Debtors.* Debtors are stated after deduction of specific provisions for debts considered bad or doubtful.

(i) *Taxation.* Taxation is charged at the current rate on the profit of the year. Provision is made for deferred taxation using the liability method in respect of future tax recoveries or charges where it can be demonstrated with reasonable probability that they will occur in the foreseeable future.

(j) *Foreign Currencies.* Assets and liabilities expressed in currencies other than naira have been translated to naira as follows:

(1) *Pre-1984 Items.* Liabilities existing at December 31, 1983 and denominated in US$ have been valued per the Central Bank o Nigeria Guidelines on refinancing of outstanding debts at the rate of N1 = US $1.3326; the US$ equivalent of liabilities in other foreign currencies has been arrived at by using the applicable cross rates to the US$ at the balance sheet date.

(2) All other items have been valued at the rates of exchange ruling at the balance sheet date. Gains or losses on such translation are taken to Profit and Loss Account.

Notes to the Accounts

(All amounts in 000s)	Notes	*1988* (in 000's)	*1987* (in 000's)
1. GROUP TRADING PROFIT			
is stated after charging:			
Directors' emoluments Fees		23	23
Other remuneration		393	327
		416	350
Auditors' remuneration		65	58
Depreciation		6,194	3,800
Profit/(Loss) on disposal of fixed Assets		(973)	94
2. TAXATION			
Provided on profit of the year		23,614	16,930
Adjustment in respect of prior years		—	(1,684)
		23,614	15,246
3. EXTRAORDINARY ITEM			
Disposal of Leasehold buildings		179	675
Less taxation applicable		87	564
		92	111

4. EARNING PER STOCK
The calculation of earnings per stock is based on the profit after taxation and on the 100 million stock units issued and fully paid.

5. FIXED ASSETS	*Leasehold Properties*	*Plant and Equipments*	*Motor Vehicles*	*1988 Total*	*1987 Total*
Cost or valuation:					
At October 1, 1987	85,354	38,910	4,440	128,704	95,219
Revaluation surplus/(application to disposal)	(2,186)	—	—	(2,186)	33,018
Additions during the year	3,103	15,037	610	18,750	5,401
Value of disposals	(149)	(4,004)	(295)	(4,448)	(4,937)
At September 30, 1988	86,122	49,943	4,755	140,820	128,704
Depreciation:					
At October 1, 1987	—	19,388	2,228	21,616	24,017
Charge for the year	2,345	2,969	880	6,194	3,800
Transfer to capital Reserve	—	—	—	—	(3,094)

	Leasehold Properties	Plant and Equipments	Motor Vehicles	1988 Total	1987 Total
Relating to Disposals	(29)	(2,780)	(133)	(2,942)	(3,107)
At September 30, 1988	2,316	19,577	2,975	24,868	21,610
Net Book Value: At September 30, 1988	83,806	30,366	1,780	115,952	107,088

Leasehold Properties comprise the following:	Cost or Valuation	Depreciation	Net	Net
Long-term Leases	45,371	502	44,869	44,969
Short-term Leases	40,751	1,814	38,937	40,394
	86,122	2,316	83,806	85,354

All Company Leaseshold properties were revalued at September 30, 1987 on the basis of open market valuation, or in locations where pure investment market is nonexistent, or depreciation replacement cost.

		Group		Company	
		1988	1987	1988	1987
	Note	000's	000's	000's	000's
6. INVESTMENTS (at cost)					
Subsidiary Company		—	—	1,300	1,300
Others		228	228	—	—
		228	228	1,300	1,300
7. STOCKS					
Raw Materials		43,040	31,591	43,040	31,591
Finished goods		2,438	2,493	2,438	2,493
Other stocks		91,348	54,947	91,348	54,947
		136,826	89,031	136,826	89,031
8. CREDITORS					
Included in creditors are dividends claimable by shareholders entitled thereto, within the statutory period of limitation, amounting at		1,853	1,428	1,853	1,428

		Group		Company	
		1988	*1987*	*1988*	*1987*
	Note	*000's*	*000's*	*000's*	*000's*
9. AMOUNTS DUE TO GROUP COMPANIES					
Subsidiary Company		—	—	1,983	1,984
Holding Company and fellow Subsidiaries: for					
goods & services		13,301	11,443	13,301	11,443
for dividends not remitted		7,071	5,743	7,071	5,743
		20,372	17,186	22,355	19,170

10. SHARE CAPITAL

				1988	*1987*
Authorized—100,000,000 stock units of 50k each				50,000	25,000
Issued and fully paid—100,000,000 stock units at 50k each				50,000	25,000

(There was 1 for 1 bonus issue during the year)

	Capital	Fixed Assets and Stock Inflation	Other Revenue	Total
11. RESERVES GROUP				
At October 1, 1987	72,941	32,750	22,399	128,090
Appropriation	—	6,750	2,659	9,409
Capitalisation (Bonus issue)	—	(12,5000)	(12,500)	(25,000)
Relating to disposal of Leasehold Property	(2,186)	—	—	(2,186)
	70,755	27,000	12,558	110,313
COMPANY				
At October 1, 1987	72,800	32,750	21,620	127,170
Appropriation	—	6,750	2,657	9,407
Capitalization (Bonus issue)	—	(12,500)	(12,500)	(25,000)
Relating to Disposal of Leaseshold Property	(2,186)	—	—	(2,186)
	70,614	27,000	11,777	109,391

12. DEFERRED TAXATION

Note 5 shows the total net book value of the company's fixed assets for accounting purposes, which amount exceeds the total written down value of those assets for tax purposes by N91.360 million (1987-N93.184 million). This excess allowed for income tax purposes and also because income tax capital allowances exceed depreciation in respect of certain assets.

No provision has been made in the accounts for deferred taxation, which at the current rates of income tax and capital gains tax, would amount to N23,027 million (1987—N23.320 million).

		1988	1987
13.	CAPITAL COMMITMENTS		
	At September 30, 1988 the Directors have authorized future capital expenditure for which no provision has been made in these accounts.		
	The amounts are approximately:		
	Contracted	3,343	2,819
	Not Contracted	835	590
		4,178	3,409
14.	CONTINGENT LIABILITIES		
	There are contingents liabilities in respect of guarantees given for bank loans to farmers and staff, amounting to	2,927	3,074

15. The accounts were approved by the Board at a meeting held on November 8, 1988.

State of Source and Application of Funds for the year ended September 30, 1988

	1988 (in 000's)	1987 (in 000's)
SOURCE OF FUNDS FROM OPERATIONS		
Profit before taxation	59,931	42,900
Extraordinary item	179	675
	60,110	43,575
ADJUSTMENT FOR ITEMS NOT INVOLVING MOVEMENT OF FUNDS		
Depreciation	6,194	3,800
Prior year adjustment	—	43
Disposal of Assets	1,506	1,830
	67,810	49,248
APPLICATION OF FUNDS		
Purchase of Fixed Assets	18,750	5,404
Dividend paid	13,500	10,000
Income Tax Paid	15,836	11,201
	48,086	26,605

	1988 (in 000's)	*1987* (in 000's)
CHANGES IN WORKING CAPITAL		
Stock	47,795	46,136
Debtors	(3,916)	(2,372)
Creditors	(10,374)	2,557
Deposits (Imports & Remittance)	(5.856)	4,223
Liquid Funds	(7,925)	(27,900)
	19,724	22,643
	67,810	49,248

Five years at a Glance
(All amounts in 000's)

	1988	*1987*	*1986*	*1985*	*1984*
TURNOVER	190,475	133,627	107,458	89,365	79,715
GROUP PROFITS					
Before taxation	59,931	42,900	31,348	24,118	16,020
After taxation	36,317	27,654	17,248	13,845	8,730
Extraordinary item	92	111	—	—	—
Dividends	27,000	13,500	10,000	8,250	5,000
Retained Profit	9,409	14,265	7,248	5,595	3,730
Earnings per 50k stock unit (adjusted)	36.3k	27.7k	17.2k	13.8k	8.7k
Dividends per 50k stock unit (adjusted)	27.0k	13.5k	10.0k	8.2k	5.0k
GROUP BALANCE SHEET					
Assets Employed:					
Fixed Assets	140,820	128,704	95,219	92,102	90,585
Depreciation	24,868	21,616	24,017	20,400	17,436
Net fixed assets	115,952	107,088	71,202	71,702	73,149
Goodwill and trade marks	—	—	—	—	3,000
Investments	228	228	228	228	228
Current Assets	159,059	128,961	108,875	84,601	66,764
Total Assets	275,239	236,277	180,305	156,531	143,141
Funds Employed:					
Share capital	50,000	25,000	25,000	25,000	25,000
Reserves	110,313	128,090	77,670	70,867	68,272
Stockholders' funds	160,313	153,090	102,670	95,867	93,272
Bank loans	—	—	—	1,800	3,000
Current liabilities	114,926	83,187	77,635	58,864	46,869
	275,239	236,277	180,305	156,531	143,141

The earnings and dividends per stock unit have been calculated on the basis of the issued stock at September 30, 1988.

Statement of Value Added

	000's	%
Sales to Third Parties	247,395	100.0
Less: Bought in Materials & Services	96,413	39.0
Nigerian suppliers for goods and services	36,721	14.8
Nigerian farmers for tobacco input	22,212	9.0
Overseas suppliers for goods and services	37,480	15.2
Value Added (wealth created by the Company	150,982	61.0
Government:		
Company Tax	23,701	15.7
Excise on Manufacture	56,920	37.7
Employee's Salaries, Wages and Allowances	27,758	18.4
Shareholders—Dividends Payable	27,000	17.9
Provide for Replacement of Assets and		
Growth	15,603	10.3
	150,982	100.00

Statement of Transactions in Foreign Currency

	Cash	Accrual
Operating Flows	(in 000's)	(in 000's)
Export of Goods/Services	563	563
Import of Raw Materials/Services	(2,319)	(13,480)
	(1,756)	(12,917)
Other Export Revenue	—	—
Other Import Expenses	—	—
Total Net Operating Flows	(1,756)	(12,917)
Capital Flows		
Inflows	—	—
Outflows	—	—
Total Foreign Export Output	(1,756)	(12,917)

REFERENCES

Accounting Standards Steering Commitee. 1975. *The Corporate Report*. London: ASSC.

Hogg, R.V., and E. Tanis. 1983. *Probability and Statistical Inference*. New York: Macmillan.

Meek, G.K., and S.J. Gray. 1988. The Value Added Statement: An Innovation for U.S. Companies. *Accounting Horizons* (June): 73-81.

Nigerian Accounting Standards Board. 1984. *Disclosures in Published Financial statements: Statement of Accounting Standard No 2*. Lagos: NASB.

PART III

INTERNATIONAL MANAGEMENT ACCOUNTING

DIVERSIFICATION STRATEGY AND ECONOMIC PERFORMANCE OF FRENCH FIRMS

Ahmed Riahi-Belkaoui

ABSTRACT

This study examines the nature of the relationships between diversification strategies and economic performance of French firms. The results show that the diversification performance relationship was different for accounting measures than for market-based measures. Unrelated diversifiers had a significant negative performance based on accounting measures. Related diversifiers had a significant positive performance based on market-based measures.

INTRODUCTION

This paper examines the relationship between business diversification strategy and corporate performance of large multimarket French firms. The business

Advances in International Accounting,
Volume 5, pages 163-171.
Copyright © 1992 by JAI Press Inc.
All rights of reproduction in any form reserved.
ISBN: 1-55938-415-8

strategies examined are either related diversification or unrelated diversification. The corporate performance measures are either accounting-based or market-based measures of performance. The thesis of this paper is that the above two distinctions are important to the diversification performance relationship in France, in that it will be different for accounting measures than for market-based measures.

RELATED RESEARCH AND HYPOTHESIS

The multidivisional form (M-form) structure is a popular device for multinationals in response to the problems of managing growth and diversity within a centralized (U-form) structure. It serves to overcome the problems of both internal and strategic control that confront large multiproduct firms (Williamson 1975; Belkaoui 1986, 1991). In addition, firms resort to two main strategies to become multiproduct enterprises: related diversification and unrelated diversification. Synergistic economies are realized by firms that have diversified into a related set of businesses, that is, related diversification strategy. Financial economies are realized by firms that have diversified into unrelated areas, that is, unrelated diversification strategy. A goal of international strategy research is to show that the multidivisional structure creates performance levels in firms that employ the different strategic approaches of unrelated versus related diversification (Cable and Yasuki 1985; Kim, Hwang, and Burgers 1989; Keats 1990). Using various measures of diversification a steam of strategy research showed in the U.S. context findings of two kinds: either (1) the related diversifiers outperformed unrelated diversifiers (Rumelt 1974; Bettis 1981; Palepu 1985; Varadarajan and Ramanujan 1987), or (2) the unrelated diversifiers outperformed the related diversifiers (Michel and Shaked 1984; Dubofsky and Varadarajan 1987). The differences in the results are due to the failure to (1) use both accounting and market-based measures of performance, and (2) consider the role of time in the causal ordering of diversification performance outcome relationships. By correcting for these two limitations, Keats (1990) showed in the U.S. context that (1) related diversification exhibited significant relationships with both dimensions of performance, and (2) unrelated diversification exhibited significant relationships with only the market-based measures of performance. Basically, both diversification and performance are multidimensional constructs and identification of appropriate criteria for performance assessment depends on the strategy pursued.

This study relies on the same strategy of (1) using both accounting and market-based measures of performance, and (2) using diversification data that precede the period for which performance data are collected, to argue that the diversification-performance relationship in France should also be different for accounting measures than for market measures. Basically, the thesis is that French related diversifiers are capable of efficiently exploiting the economies of scope that would be viewed favorably by investors as they may feel that they can diversify their portfolios more efficiently than a firm can (Amihud and Lev 1981). In addition, the failure of most unrelated diversifiers to act as true holding companies with an efficiently working internal capital market, may result in underperformance as measured by accounting-based measures of performance. Accordingly, the following needs to be tested.

Hypothesis 1. Related diversifiers in France will demonstrate positive performance based on market-based measures of performance. Unrelated diversifiers in France will demonstrate negative performance based on accounting-based measures of performance.

Basically, French firms were chosen given that empirical studies have already found evidence of substantial M-form gains in America and Britain (e.g., Armour and Teece 1978; Steer and Cable 1978; Thompson 1981) and only short-run costs and no evidence of eventual long-run gains in West Germany (Cable and Dirrheimer 1983) and in Japan (Cable and Yasuki 1985). Empirical evidence on the French situation for both the multiproduct hypothesis and the diversification performance hypothesis is seriously lacking.

METHODS

Sample of French Firms

The firms included in the present study were taken from the tenth annual edition of the *French Company Handbook* (International Business Development and International Herald Tribune 1989/1990). The handbook includes accounting, market, and structural information for a selection of French firms. The companies selected to be included in the handbook are identified as the most important French firms. Of the 83 firms listed, 52 met the criteria for inclusion in this study: primarily the availability of data needed for the computation of both performance and diversification measures. Government owned firms were excluded. A list of the firms studied is shown in the Appendix.

Performance Measures

The measures of performance used were either accounting or market-based as established in international strategy research (Keats 1990; Cable and Yasuki 1985). They include two accounting measures and two market-based measures.

The two accounting measures are the rate of return on assets and the undiluted earnings per share for the year 1988. The two market-based measures are the 1988 price earnings ratio and the 1988 Tobin Q computed as the 1988 market value of the firm divided by the 1988 historical costs of the assets of the firms. Values for all these measures are obtained from the 1989 and the 1990 *French Company Handbook.*

Diversification Measures

The choice of a diversification measure is crucial to the empirical investigation of performance implications of corporate diversification strategy. Given Montgomery's (1982) results (1982) on the strengths and weaknesses of the index approaches used by industrial organizational researchers and the categorical approach used by strategy researchers, Palepu (1985) relied successfully on the Jacquemin-Berry (1979) measure. This entropy measure is used in this study. Its computation follows.

Given a firm operating in N industry segments, the entropy measure of total diversification DT is defined as follows:

$$DT = \sum_{i=1}^{n} P_i \, \mathrm{Ln}\,(1/P_i)$$

where P_i is the share of the ith segment in the total sales of the firm. Ln is the logarithm.

The related diversification, DR_j is defined as follows:

$$DR_j = \sum_{i \in j} P_i^j \, \mathrm{Ln}\,(1/P_i^j)$$

where P_i is defined as the share of the ith segment of group j in the total sales of the group.

The total related diversification DR is a function of DR_j. It is defined as follows:

$$DR = \sum_{j=1} DR_j \, P_j$$

where P_j is the share of the jth group sales in the total sales of the firm.

The unrelated diversification DU is computed as follows:

$$DU = \sum_{j=1}^{m} P^j \, \mathrm{Ln}\,(1/P^j)$$

Values of *DR* and *DU* are computed from data derived from the 1989 *French Company Handbook.*

RESULTS

Table 1 shows the results of the regression between the 1988 accounting-based and market-based measures derived from the 1990 *French Company Handbook* and the 1987 measures of related and unrelated diversification derived from the 1989 *French Company Handbook.* The results on the *F*-statistic reveal a significant relationship between the measures of performance and the measures of diversification for the French companies examined. The second significant result is that the diversification performance relationship is different for accounting measures than for market-based measures. Basically, both accounting-based measures of performance, rate of return on assets, and earnings per share are negatively related to the level of unrelated diversification, while both market-based measures of performance, price earnings ratio, and Tobin Q are positively related to the level of related diversification.

DISCUSSION

These results on French company data show that diversification and performance are multidimensional phenomena and the appropriate performance evaluation criterion depends on the strategy used. First, the related business diversification of the French firms exhibited a strong positive link with market-based performance measures. This suggests that in the French context, investors value a related diversification strategy more than an unrelated diversification one. This points to the necessity of using market-based measures in evaluating the related diversification strategies of French firms.

Related diversifiers are best organized to exploit economies of scope or the sharing of resources and capabilities among a related set of businesses (Teece 1982). This capability is highly valued by investors in the French market. The failure of accounting-based measures to reflect related diversification is due to the failure of accounting data to account for interdivisional coordinations among divisions in related diversifiers. As Porter suggested: "[i]nterrelationships almost inevitably introduce some subjectivity into performance measurement...because business unit contributions to the firm as a whole are often hard to quantify precisely" (1985, 392). As a result, in such a case Kerr (1985) calls for the use of a mix of objective and subjective criteria, and Hoskisson (1987) suggests that in related diversifiers resource allocation based on objective financial performance criteria is compromised. The results for the French study confirms the inappropriateness of the

Table 1.   Regression Results

Dependent Variables	Independent Variables[a]			R^2 (%)	F
	Intercept	Related Diversification	Unrelated Diversification		
Rate of Return on Assets	0.0809	−0.01146	−0.0513		
	(5.34)*	(1.07)	(2.36)**	12.37	3.25**
Earnings Per Share	31.4706	−9.2575	−21.799		
	(5.24)*	(−1.06)	(−2.54)**	19.48	4.96*
Price/Earnings	6.1797	16.015	3.2048		
	(0.59)	(1.89)**	(0.21)	8.02	3.79**
Tobin Q	0.3239	1.131	−0.148		
	(0.41)	(1.90)***	(−0.13)	8.36	3.81**

Notes: [a] *t*-statistics are in parentheses
           * significant at $\alpha = 0.01$
          ** significant at $\alpha = 0.05$
         *** significant at $\alpha = 0.10$

accounting-based measures of performance (Rumelt 1974; Bettis 1981; Palepu 1985; Varadarajan and Ramananujan 1987; Dubofsky and Varadarajan 1987).

To realize the financial advantages of a strategy of unrelated diversification, both decomposition of the firm into distinct divisions and decentralization of operating responsibilities to those divisions are needed. The underperformance of unrelated French diversifiers in this study may be due to the large focus on integrative effort that may compromise divisional autonomy and accountability, thereby reducing the effectiveness of any existing internal capital market (Lorsch and Allen 1973; Dundas and Richardson 1982). Unrelated French diversifiers would be more profitable if they were operated as holding companies.

The results of this study call for more empirical and theoretical research on both the diversification performance relationship issue and the role of accounting data versus market data in performance evaluation internationally. First, with regards to the diversification performance relationship issue, the results of this study in the French context call for replication using different accounting and market-based measures of performance and different measures of diversification, as well as different periods. In addition, evidence is needed in other European and developing countries' contexts, especially in those countries where a "thin" capital market may make the market-based measures of performance less reliable.

Second, with regard to the role of accounting data versus market data in performance evaluation internationally, this study highlighted the crucial importance of the moderating effects of diversification strategy. Future research may explore the moderating effects of a host of variables. Examples include the effects of ownership structure, market concentration, industrial sector, and so forth. The effects of some of these variables on performance evaluation need to be explored in various countries, developed and developing, before a general and workable "theory of performance evaluation internationally" may be developed.

APPENDIX: LIST OF FRENCH FIRMS

1. Accor
2. Beghin-Say Group
3. B S N
4. Cap Gemini Sogeti
5. Ceglelec
6. C.E.P. Communication
7. C.G.I.P.
8. C.G.M. Group
9. Chargeurs S.A.
10. Club Mediterraneè
11. CMB Packaging
12. Companie Financière de Suez
13. Companie Gènèrale d'Electrecitè
14. Companie Gènèrale des Eux
15. CRP-Compagnie Parisienne de Rèescompte
16. Dassault
17. Docks de France
18. Electronique Serge Dassault
19. Elf Auitaine
20. Epèda-Bertrand Faure
21. Essilor
22. Framatome
23. Havas
24. Imetal
25. Lafarge Coppèe
26. Legrand
27. L'Orèal
28. LVMH Moët Hennessy Louis Vuitton
29. Lyonnaise des Eaux
30. Michelin
31. Moulinex
32. Pechiney
33. Pernod Ricard
34. Peugeot S.A.

35. Poliet
36. Printemps Group
37. La Redonte
38. Rhône-Poulenc
39. Saint-Gobain
40. Salomon
41. Sanofi
42. SCOR S.A.
43. SEB Group
44. Seita
45. Sema Group
46. SGE Group-Sociètè Gènèrale d'Enterprises
47. Skis Rossignol
48. Sligos
49. Sociètè Gènèrale
50. SODEXHO
51. Sommer Allibert
52. Thomson
53. TOTAL
54. Usinor Sacilor
55. Valeo
56. Vallourec
57. Vicorie Sacilor

REFERENCES

Amihud, Y., and B. Lev. 1981. Risk reduction as a managerial motive for conglomerate mergers. *Bell Journal of Economics* 12: 605-17.

Armour, H.O., and D.J. Teece, 1978. Organizational structure and economic performance: A test of the multidivisional hypothesis. *Bell Journal of Economics* 9: 106-22.

Belkaoui, A. 1986. *Management Control Systems.* Westport, CT: Quorum.

———. 1991. *Multidimensional Management Accounting Systems.* Westport, CT: Quorum.

Bettis, R.A. 1981. Performance differences in related and unrelated firms. *Strategic Management Journal* 2: 379-93.

Cable, J.R., and H. Yasuki. 1985. International organization, business groups and corporate performance. *International Journal of Industrial Organization* 3: 401-20.

Cable, J.R., and M.J. Dirrheimer. 1983. Hierarchies and markets: An empirical test of the multidimensional hypothesis in West Germany. *International Journal of Industrial Organization* 1: 43-62.

Dubofsky, P., and P. Varadarajan. 1987. Diversification and measures of performance: Additional empirical evidence. *Academy of Management Journal* 30: 597-608.

Hoskisson, R.E. 1987. Multidivisional structure and performance: The contingency of diversification strategy. *Academy of Management Journal* (December): 625-44.

International Business Development and International Herald Tribune. 1989/1990. *French Company Handbook 1990.* Paris: French Company Handbook.

Jacquemin, A.P., and C.H. Berry. 1979. Entropy measure of diversification and corporate growth. *The Journal of Industrial Economics* (June): 359-69.

Keats, B.W. 1990. The vertical construct validity of selected business economic performance indicators. *Journal of Applied Behavioral Science* 29(2): 151-60.

Kerr, J.L. 1985. Diversification strategy and managerial rewards: An empirical study. *Academy of Management Journal* 28: 155-79.

Kim, W.C., P. Hwang, and W.P. Burgers. 1989. Global diversification strategy and corporate profit performance. *Strategic Management Journal* 10: 45-57.

Lorsch, J.W., and S.A. Allen. 1973. *Managing Diversity and Interdependence.* Boston: Division of Research, Harvard Graduate School of Business Administration.

Michel, A., and I. Shaked. 1984. Does business diversification affect performance? *Financial Management* 4: 18-25.

Montgomery, C. 1982. The measurement of firm diversification: Some new empirical evidence. *Academy of Management Journal* 25: 299-307.

Palepu, K. 1985. Diversification strategy, profit performance and the entropy measure. *Strategic Management Journal* 6: 239-55.

Porter, M. 1985. *Competitive Advantage: Creating and Sustaining Superior Performance.* New York: Free Press.

Rumelt, R. 1974. *Strategy, Structure and Economic Performance.* Cambridge: Harvard University Press.

Steer, P.S., and J.R. Cable. 1978. Internal organization and profit: An empirical analysis of large U.K. companies. *Journal of Industrial Economics* 27: 13-30.

Teece, D.J. 1982. Towards an economic theory of the multiproduct firm. *Journal of Economic Behavior and Organization* 3: 39-63.

Thompson, S. 1981. Internal organization and profit: A note. *Journal of Industrial Economics* 1: 30.

Varadarajan, P., and V. Ramanujan. 1987. Diversification and performance: A re-examination using a new two-dimensional conceptualization of diversity in firms. *Academy of Management Journal* 30: 380-93.

Williamson, O.W. 1975. *Markets and Herarchies: Analysis and Antitrust Implications.* New York: Free Press.

ORGANIZATIONAL AND INTERNATIONAL FACTORS AFFECTING MULTINATIONAL TRANSFER PRICING

Susan C. Borkowski

ABSTRACT

International transfers between a parent multinational corporation (MNC) and its subsidiaries must be assigned a price if the MNC uses responsibility center accounting. While domestic transfer prices are driven by the profit maximization criterion, international transfers are also affected by tax and tariff regulations, currency stability, and other nondomestic concerns. Like their domestic counterparts, the majority of U.S.-based MNCs use a cost-based transfer price rather than the theoretically preferred market, negotiated, or programming methods. This study found that the majority of MNCs use dual sets of books to counter the problems caused by a given transfer pricing method, contrary to prior findings. Significant organizational factors in choosing a method include ease and cost of implementation, the use of subsidiary profit as the primary

Advances in International Accounting,
Volume 5, pages 173-192.
ISBN: 1-55938-415-8

performance evaluation measure, and the degree of decentralization in the MNC. Significant international factors include tax and tariff regulations, the economic stability of the parent MNC country, and Sec. 482 regulations.

Many large companies transfer goods internally, deriving some price for these goods. In multinational corporations (MNCs), this transfer can occur between a subsidiary operating in a foreign country and the parent company, or other subsidiaries in the corporation. Transfer pricing affects the reported profit of a subsidiary, whether foreign or domestic; this reported profit is normally an important criterion in the performance evaluations of profit center managers.

The motives governing the choice of a transfer pricing method are therefore of concern to MNC subsidiary managers. External groups are also interested in and affected by the ramifications of international transfer pricing; these groups include the IRS, customs officials, antitrust officials, creditors, investors, and labor unions (Arpan 1972).

Many studies on transfer pricing are concerned with domestic applications, or with international transfer pricing primarily as a function of tax minimization. The findings of an earlier study limited to domestic transfers (Borkowski 1990) indicated that transfer pricing, whether domestic or international, is still an area of interest and concern to managers.[1] Transfer pricing for MNCs addresses many aspects of motivation and intent, because it occurs in a more complex economic and political environment than that of domestic transfers. While profit maximization is usually the dominant motivating force driving the choice of a domestic transfer pricing method, several criteria may co-dominate the choice of an international transfer pricing method. In addition to profit maximization, these criteria include compliance with U.S. and non-U.S. tax and tariff regulations, subject to international constraints such as economic restrictions, governmental regulations, and the political environment (Wu and Sharp 1979); subsidiary performance evaluation; goal congruence; competitive advantage; and currency stabilization. These criteria are not mutually supportive, however. For example, if a method is chosen to give an advantage for tax purposes, then the use of the results of subsidiary operations for performance evaluation of managers may yield unfair results, causing managerial conflict and morale problems. This potential conflict between the transfer pricing method chosen and a fair performance evaluation is a primary concern in the MNC.

This study investigates the motivational criteria driving U.S.-based MNCs to choose an international transfer pricing method. Other MNCs are not included because their tax and tariff constraints will vary by parent country. The findings of this study may lead to a resolution of conflict between managers in MNCs, and to a better understanding and analysis of the policy issues underlying international transfer pricing choice.

An overview of general transfer pricing theory and prior international transfer pricing[2] research are presented first. Then the framework and variables of interest are described, after which sample selection and data collection are discussed. The results of the data analysis are presented in the following section, with conclusions and implications of the findings presented in the final section.

OVERVIEW OF TRANSFER PRICING THEORY AND RESEARCH

Horngren and Foster (1987, 835) define transfer pricing as the "price charged by one segment of an organization for a product or service that it supplies to another segment of the same organization." This price is usually set using one of five approaches: economic models (market and marginal pricing), negotiated prices, mathematical programming models, behavioral models which apply the prior three methods in a social framework (Watson and Baumler 1975), and cost-based methods.[3]

International versus Domestic Transfer Pricing Theory

The results of recent empirical research in domestic transfer pricing (Price Waterhouse 1984; Eccles 1985; Borkowski 1990) show that, in a significant number of cases, the transfer pricing methods used in practice are not the methods which have theoretical support in the accounting and economic literature. This divergence is partially explained by the differing motivational criteria of companies in choosing a transfer pricing method, including profit maximization, performance evaluation of divisions, goal congruence, and the ease of understanding and the cost of administration of the method chosen. Generally, one method is chosen for domestic transfers in order to maximize profit, and another for international transfers to comply with tax and tariff regulations.

Companies making domestic transfers are free to use any method to derive the transfer price. U.S.-based MNCs are not afforded this luxury, however, for international transfers. Section 482(e) of the Internal Revenue Code (IRC) was adopted to prevent the shifting of profits between parent and subsidiary to evade U.S. taxes, and to promote the proper reflection of taxable income. The regulations pertaining to this code section stipulate that an arm's length price must be determined using one of four methods, in the following order of consideration.

1. Comparable uncontrolled price (CUP). The market price of the transferred good must be used if a market exists for the good. This method assumes an "as if unrelated" transfer between MNC and subsidiary. If no market exists, then use

2. Resale price (the selling price of the final good minus the "appropriate mark-up" on sale). If further processing or manufacturing is required on the product, then use

3. Cost plus. This is the cost of the unfinished transferred product plus the "appropriate mark-up" on cost. If none of these three methods is applicable, then use

4. Some other appropriate method, commonly called "the fourth method." Appropriate methods include basic arm's-length or reasonable rate of return, functional analysis and reasonable profit split. If an MNC feels that the other three methods are not applicable, it may choose what the Internal Revenue Service (IRS) terms any other reasonable method, but the burden of proof of its appropriateness rests with the MNC.

Each method must be examined and rejected in turn before the next method is considered. Managers prefer a reordering, with cost plus as the method of choice, then comparative uncontrolled price and resale price (Burns 1980). The choice of cost plus mirrors the preferred practice in domestic transfers, a practice which has little support in the accounting literature. A recent IRS White Paper (U.S. Department of the Treasury and Internal Revenue Service 1988) was expected to deal with reordering; instead, its focus was on the pricing of intangibles. The White Paper did contain suggestions for changing the current regulations pertaining to IRC Section 482. The comparable uncontrolled price would still be the preferred method; if not applicable, then MNCs may choose from the other three methods without prejudice.

In practice, however, no method is optimal; the best method is contingent upon each MNC's specific situation. For example, when the tax rates differ between the United States and the subsidiary country, the resale method and cost plus method (favored by managers) "cause resource allocation distortions or inefficiencies: that is, they cause the MNE to optimally allocate resources differently than it would in the absence of taxation" (Halperin and Srinidhi 1987, 687). The authors of the White Paper recognize this contingent nature of transfer pricing choice by recommending penalties only in the case of obvious manipulation by an MNC to avoid taxes, which would moderate current IRC Section 482 penalties.

Schindler (1988) attributes the tax problems with transfer pricing to IRC Section 482 itself, with its arbitrary ranking of "acceptable" methods, the lack of comparables to establish an arm's-length price, and the post-choice burden of proof on the MNC. One solution is the use of (and the IRS' full acceptance of) the functional analysis/profit split approach as the fourth method, concurrent with an updating of IRC Section 482 to reflect the realities of an MNC's operating environment.

The IRS is now granting preapproval of the method chosen for international transfers. Due to the increasing number and cost of cases in litigation regarding an MNC's choice of method, the IRS thinks that it will be cost-effective to work with MNCs in preapproving transfer pricing methods. This advanced ruling process would mitigate the current situation where the MNC chooses a method and then is required, if audited, to justify that choice to the IRS.

Given IRC Section 482 constraints, however, it is difficult for an MNC to sufficiently manipulate prices to achieve the ideal maximization of both reported costs in the country with the higher tax rate, and reported revenues in the country with the lower tax rate. Brooke and Remmers (1972, 275-76) feel the

> burden of proof is put on the company to demonstrate that the method of transfer pricing is reasonable.... It is less useful than is often believed as a means to avoid taxes, particularly for U.S. companies. This is because in order to produce any real tax savings, there would have to be an optimum combination of substantial tax differentials between countries, adequate profit margins, and perhaps low custom duties.

Most research in MNC transfer pricing has found that stated company goals and objectives conflict. External considerations include minimization of (1) income tax, (2) import/export duties, (3) the impact of exchange rate fluctuations on evaluation of divisions and on tax minimization, and (4) non-U.S. government influence. These may conflict with internal considerations such as (1) evaluation of divisions using budgets, where the parent judges the subsidiary, but also influences or sets the transfer price, affecting the budget, (2) discouraging competitors from entering the market by setting an artificially high transfer price to show misleading false low profits, or (3) gaining a competitive advantage in a new or existing market by setting an artificially low transfer price (Coburn, Ellis, and Milano 1981). Problems include the conflict between corporate goals and subsidiary morale, little freedom in choosing the transfer price when operating in certain restrictive countries, and the costs of transfer price manipulations most likely exceeding the benefits received. If the parent MNC dictates the transfer price, then subsidiary autonomy and participation are limited, as is the managers' ability to respond to changes in local economic and political conditions.

Knowles and Mathur (1985) also discuss the transfer pricing objectives of MNCs, such as augmenting benefits via goal congruence, evaluation and motivation, and restricting costs. These objectives are broad and seemingly complementary. However, they and other researchers find that the functional aspects of these objectives are not as complementary, because they include the simultaneous but not mutually inclusive goals listed in Table 1. The conflicts

Table 1.   Factors in Determining International
Transfer Pricing Methods

Overall profit to parent company	Most authors
Goal congruence	Most authors
Performance evaluation of subsidiary	Most authors
Minimize parent company's overall tax burden	Brook and Remmers (1972) Arpan (1972) Tang (1979, 1982) Coburn et al. (1981) Knowles and Mathur (1985)
Provide competitive advantage to subsdiary in other country	Brook and Remmers (1972) Arpan (1972) Tang (1979, 1982) Coburn et al. (1981) Knowles and Mathur (1985)
Limit exchange losses due to currecy de- or re-evaluation	Brook and Remmers (1972) Arpan (1972) Coburn et al. (1981) Knowles and Mathur (1985)
Avoid restrictions on repatriation of profits	Arpan (1972) Tang (1979, 1982) Knowles and Mathur (1985)
Understate subsidiary profit to avoid local price reduction or wage increase	Brook and Remmers (1972) Knowles and Mathur (1985)
Limit custom duties	Arpan (1972) Coburn et al. (1981) Knowles and Mathur (1985)
Avoid sharing profit with foreign partners in joint ventures	Brook and Remmers (1972) Knowles and Mathur (1985)
Discourage competitors from entering the market	Coburn et al. (1981)

among maximizing profit, minimizing taxes and tariffs, and providing ;
meaningful basis for performance evaluation exist in most MNCs. The physica
distances between parent and subsidiary, coupled with tax differentials, caus
many MNCs to centralize more decisions, causing difficulty in measuring th
subsidiary's performance.

To deal with conflicting goals, Brooke and Remmers (1972), Coburn et al
(1981), and Tang (1982) found varying approaches in practice by MNCs
including:

1. accounting for transfer price manipulations in the budget. This is th
    prevalent practice among MNCs, but causes a problem with motivatio:

when managers are evaluated against planned results, even if a loss is the planned result.

2. using two sets of books (one for tax, finance and local purposes, and the other for management control purposes). Disadvantages include cost, complexity, and difficulty in measuring the costs and benefits of maintaining an expensive and cumbersome dual system.

3. approximating in the subsidiary those conditions that would be faced by an independent market entity. This approach does not distort profits, but disregards tax goals and makes evaluation difficult.

4. disregarding the effects of transfer prices. A problem can arise with job satisfaction, which is partly derived from achieving what managers consider reasonable profits in absolute terms. If profit is made, how can a manager determine what part is due to subsidiary efficiency, what part is due to favorable transfer prices, and is the profit satisfactory?

Findings of Prior Surveys

Early studies identified some factors affecting MNCs when choosing a transfer price. Cost-based methods were preferred, and profit maximization concurrent with tax minimization were the MNCs' primary objectives. A summary of empirical studies on international transfer pricing is presented in Table 2.[4]

Tang (1979) discovered that cost-based prices were preferred by the majority of U.S.-based MNCs. The primary factor affecting method choice was profit maximization, followed by avoiding restrictions on repatriation of profits, competitive advantage, and income tax minimization. The chief objectives of MNCs were profit maximization and subsidiary performance evaluation. The transfer pricing decision was centralized, made by the parent MNC. No size relationships were found.

Wu and Sharp (1979) found that market prices were preferred by MNCs when an external price was available, followed by full cost. Full cost was preferred when a market price did not exist, with negotiation as a second choice. MNCs were not asked which method they actually used, but how important a method is in a specific situation. Important motivational criteria included compliance with non-U.S. tax and tariff rules, followed by profit maximization and U.S. tax considerations.

Burns (1980) restricted her study to the effects of IRC Section 482 on U.S.-based MNCs. Managers disagreed with the required order of the methods (CUP, resale, cost-plus, and other), preferring instead cost-plus, followed by CUP and resale, with little need for a fourth method. She concluded that the problem with IRC Section 482 is that it is based on the "premise that a

Table 2.   Prior Empiricial Studies on International Transfer Pricing

Study	Methodology	Respondents
Business International Corporation (1965)	Interview	30 MNCs
Shulman (1966)	Interview	8 U.S. MNCs
Bisat (1966)	Questionnaire	14 international CPA firms
Conference Board (1970) (Green and Duerr)	Questionnaire	130 U.S. MNCs
Arpan (1972)	Questionnaire	60 non-U.S. MNCs
Business International Corporation (1973)	Interview	MNCs
Milburn (1977)	Questionnaire and interview	33 Canadian/U.S. accounting firm partners
Tang (1979)	Questionnaire	105 U.S. and 75 Japanese MNCs
Wu and Sharp (1979)	Questionnaire	61 U.S. firms
Burns (1980)	Questionnaire	62 U.S. MNCs (Section 482 audits)
Tang (1982)	Questionnaire	78 Canadian and 48 British MNCs added to 1979 study
Yunker (1983)	Questionnaire	52 U.S. MNCs
Schindler and Henderson (1985)	Questionnaire	U.S. MNCs (Section 482 audits)
Benvignati (1985)	U.S. FTC data	674 U.S. MNC lines of business
Al-Eryani, Alam, and Akhter (1990)	Questionnaire	164 U.S. MNCs

Note: See Tang (1979) for a detailed discussion of the first seven studies.

subsidiary is legally and economically separate from its parent corporation";
however, 59% of the responding MNCs do not operate in this manner (1980,
314).

In an extension of his 1979 study, Tang (1982) found that many firms use
a form of ROI as the basic performance measure for divisions, which can cause
problems with subsidiary profit measurement, performance evaluation, and
decentralization of decision making.

Yunker (1983) found that larger MNCs tended toward market-based transfer
prices, and were more likely to use profit-oriented measures in evaluating
subsidiary performance. Cost-based firms were more concerned with budgetary
and goal-oriented performance criteria.

Schindler and Henderson (1985) further investigated IRC Section 482 audits
and concurred with previous findings that the arms-length standard mandated
by the IRS was not reflective of MNC operations, and that the regulations
under IRC Section 482 allowed the IRS to subjectively judge an MNC's method
choice and unfairly place the burden of proof on the MNC in an audit.

Benvignati (1985) used data from the U.S. Federal Trade Commission, analyzing lines of business rather than MNCs. She found that the majority (57%) used cost plus, followed by either comparable uncontrolled price or resale (24%), or some fourth method (19%).

Recently, Al-Eryani et al. (1990) found that MNCs concerned with legal compliance with tax and customs regulations used market based transfer prices. Larger firms also tended toward market-based methods.

METHODOLOGY

MNCs are similar in that they all account for domestic as well as international transfers; beyond this, each operating environment is unique. The IRS, in allowing an MNC to choose its transfer pricing method, indirectly acknowledges these varied environments, and effectively supports the contingency theory approach to accounting. Each MNC develops its own different optimal set of accounting and management practices which are situation-specific and contingent upon the organizational and environmental characteristics of the MNC's operating milieu. The framework for this study is adapted from Schweikart's (1985, 1986) contingency theory-based managerial accounting model, and incorporates Birnberg and Shields' (1984) cognitive psychology model of perception and attention to information in the environment. The model for this study, presented in Figure 1, specifically addresses the MNC's process in choosing an international transfer pricing method. Organizational and international environmental variables will vary among MNCs, leading to different yet appropriate transfer pricing decisions for each MNC.

In domestic transfers, method choice is determined by a combination of organizational and environmental factors. International transfer pricing requires consideration of additional factors peculiar to intercountry transactions, such as taxes, tariffs, customs duties, and political relationships. The hypotheses under consideration are the following.

Hypothesis 1.   The choice of an international transfer pricing method is not affected by organizational (internal) variables.

Hypothesis 2.   The choice of an international transfer pricing method is not affected by international (external) variables.

The organizational and international variables are as follows:

ORG1:   Size of the MNC
ORG2:   Conflict within the MNC

ORG3:   Transfer pricing method requirements-internal
ORG4:   Performance evaluation criteria
ORG5:   Management compensation
ORG6:   Profit orientation/MNC business objectives
ORG7:   Degree of decentralization
INTL1:   Size of parent/non-U.S. subsidiary transfers
INTL2:   Industry
INTL3:   Transfer pricing method requirements-external
INTL4:   Economic stability
INTL5:   Economic favorableness
INTL6:   MNC practices to counter effects of transfer pricing
INTL7:   IRC Section 482 requirements.

The dependent variable is the transfer pricing method, which will be analyzed in two dimensions: actual method used, and optimal method given IRC Section 482 regulations.

Variable Selection

The size of the MNC (ORG1) was determined by net sales dollars. Earlier findings regarding size were contradictory, with larger MNCs associated with cost-based methods (Arpan 1972), or with market-based methods (Yunker 1983; Benvignati 1985), or with no method at all (Tang 1979).

Conflict (ORG2) was measured between parent and subsidiary managers, and among subsidiary managers, through the respondents characterizing the degree of conflict at various managerial levels using a five-point scale. Conflict can increase when transfer pricing methods affect performance criteria, or when the methods are dictated to the subsidiary by the parent MNC.

A transfer pricing method must meet both organizational (ORG3) and international (INTL3) requirements. Responses to 28 items were factor analyzed to yield four summary variables for ease and cost of implementation (ORG3a), evaluation (ORG3b), decision making (ORG3c), and international concerns (INTL3). The reduction of multiple associated variables affecting transfer pricing choice to a fewer number of interpretable factors has been successfully used by Rushinek and Rushinek (1988) and Borkowski (1990).

Performance evaluation criteria (ORG4) and compensation (ORG5) are important because MNCs address the conflicting objectives of transfer pricing by innovative accounting and reporting practices (Brooke and Remmers 1972; Coburn et al. 1981; Tang 1982). These practices may then conflict with the MNC's criteria for performance evaluation and reward. Performance evaluation criteria were factor analyzed to yield dimensions based on nonincome subsidiary measures such as sales growth and market share (ORG4a), profit ratios such as ROI (ORG4b), subsidiary income (ORG4c),

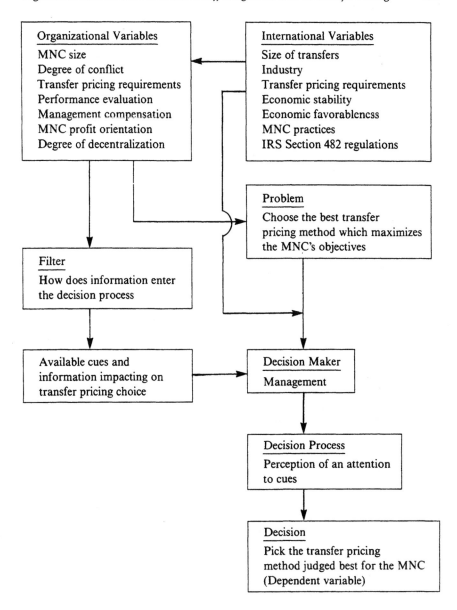

Source: Adapted from Schweikart (1985, 1986) and Birnberg and Shields (1984).

Figure 1.   Model of International Transfer Pricing
Choice within the Firm

and innovation (ORG4d). Respondents were asked to indicate whether subsidiary managers received incentive compensation in the form of a bonus, and if so, which method was used to determine the corporate bonus pool.

Profit orientation and business goals (ORG6) may affect how an MNC chooses a method and then implements it. The method chosen should be compatible with the MNC's goals.

The degree of decentralization (ORG7) affects how the method is chosen, the subsidiary's participation in decision making, and how performance is evaluated. Some items measuring this construct are derived from Vancil's (1978) study. Items were analyzed individually, and then averaged to yield an overall measure of decentralization.

The size of an MNC's international transfers (INTL1) may influence method choice. Benvignati (1985) found that MNCs with higher percentages of intracompany transfers avoided market-based transfer prices.

Findings about industry (INTL2) have been inconclusive. While most studies have found no industry effects, Wu and Sharp (1979) found a significant relationship between industry and both method choice and the importance of motivational criteria. No details are provided, however. Industry was determined from SIC codes, and collapsed into four categories reflecting process industries and nonprocess industries by homogeneity (metal), by complexity (manufacturing), and other (Price Waterhouse 1984).[5]

The stability (INTL4) and favorableness (INTL5) of both the parent MNC's and the subsidiary's economic environment should influence method choice. Market conditions, political climate, and the tenor of the subsidiary country's government may directly affect the MNC's choice. If the economy is unfavorable, the government may encourage the MNC to pick a method more favorable to the latter in order to retain their trade. Respondents were asked to assess these economic conditions on a five-point scale.

The accounting practices of an MNC (INTL6) may affect method choice, or vice versa. If performance is evaluated using profit criteria, accounting adjustments may be necessary due to distortions from the transfer pricing method chosen and currency fluctuations. Respondents indicated which of the common practices described earlier (two sets of books, approximating market conditions, disregarding transfer pricing effects, affecting the budget) were used by their MNC.

IRC Section 482 regulations (INTL7) may affect method choice. Some MNCs are less willing to risk an audit and may thus choose comparable uncontrolled price, even when another method might be better for their situation.

SAMPLE SELECTION AND DATA COLLECTION

MNCs included in the sample were drawn from the population of U.S.-based companies, and met the following criteria.

1. The MNC is listed on either the *Fortune* 500 or the *Business Week* 1000.
2. The MNC had international subsidiaries or divisions.
3. The MNC was in a manufacturing industry identified in prior studies as likely to use transfer pricing.

The final sample consisted of 301 firms meeting these criteria. A questionnaire was mailed to the person responsible for international operations, with a follow-up letter and replacement questionnaire sent to nonrespondents three weeks after the initial mailing. Of the 144 questionnaires returned (47.8% response rate), 128 were usable; 49 did not transfer products internationally, and 79 did use some transfer pricing method. No significant differences were found between early and late respondents, or between respondents and nonrespondents on the bases of industry and size.

Table 3 presents the transfer price currently used by the sample MNCs as compared with the results of earlier studies. More than half (51.9%) used some full-cost method, while 32.9% used a market-based price and 15.2% used a negotiated price.

EMPIRICAL RESULTS

Nonparametric statistics were used to analyze the variables because responses were not measured using interval or ratio scales. The Kruskal-Wallis one-way analysis of variance by ranks is used to determine whether MNCs using market, negotiated, and full cost methods are from different populations. The Wilcoxon rank-sum test (equivalent to the Mann-Whitney U-test) is used when the MNCs are aggregated into market/other or cost/other comparisons. The market/other aggregation reflects the theoretical preference for market as the method of choice, with all other methods considered nonmarket, and was most recently used by Al-Eryani et al. (1990). The cost/other aggregation is derived from the categorization used by both Benke and Edwards (1980) and by Price Waterhouse (1984). They consider negotiated pricing as a market-based technique, and contrast market and negotiated firms as one group with cost-based firms. Results of the analyses are presented by hypothesis for the following three comparisons: Mkt/Neg/Cost, Cost/Other, and Mkt/Other.

Results—Hypothesis 1

It was hypothesized that organizational (internal) variables do not affect the choice of a transfer pricing method. This first hypothesis can be rejected for several of the constructs, as shown in Table 4.

The size of an MNC (ORG1) was significant only in the Cost/Other analysis. Full cost MNCs tended to be smaller, with mean sales of $2,929 million,

Table 3. International Transfer Pricing Methods Used by
U.S.-based Multinational Companies
(In Percent)

	Multinational Transfers (n = 85) (Tang 1982)	Multinational Transfers (n = 674) (Benvignati 1985)*	Multinational Transfers (n = 79) (Current Study)
Full cost plus	32.2		46.8
Variable cost plus	1.7		0.0
Subtotal cost plus	33.9	57.2	46.8
Other full cost	10.2		5.1
Other variable cost	.8		0.0
Other cost	1.7		0.0
Subtotal other cost	12.7	14.5	5.1
Total cost-based	46.6	71.7	51.9
Market or adjusted market	34.8	24.0	32.9
Negotiated	13.6	0.0	15.2
Mathematical programming	.8	0.0	0.0
Other noncost	4.2	4.3	0.0
Non cost-based subtotal	53.4	28.3	48.1
Total	100.0	100.0	100.0

Note: * Benvignati used lines of business rather than MNCs.

compared with market MNCs ($3,621 million) and negotiated MNCs ($5,310 million).

Conflict between the parent and subsidiary (ORG2a) was less in full cost MNCs, with 59% reporting little or no conflict. This was significant in the Cost/ Other analysis, where only 34% of market and negotiated MNCs reported little or no conflict. This finding might be expected, because negotiation, by its nature, would engender conflict. Significant conflict did not exist among subsidiary managers themselves (ORG2b). This finding is expected because subsidiaries may be in different countries, leading to little interaction between managers.

Of the three factors covering organizational requirements for transfer pricing, only the ease/cost criterion (ORG3a) was significant in the Mkt/Neg/ Cost comparison. This criterion is usually a characteristic of cost methods, and was important to both cost and market MNCs. Negotiated MNCs considered evaluation criteria more important.

Of the four factors measuring performance evaluation, only the use of subsidiary net income (ORG4a) was significant in all three analyses. Ninety percent of cost MNCs felt it important, compared with 65% of market and 75% of negotiated MNCs. The insignificance of the other criteria indicate that profit is still the overriding measure of success for a subsidiary manager. Non-income measures, while not significant, were the second most important

Table 4.   Analyses by Organizational Variables Affecting
Transfer Pricing Method Choice

	Mkt/Neg/Cost		Cost/Other		Market/Other	
Variables	K-W	Signif	Wilcoxon	Signif	Wilcoxon	Signif
ORG1	2.801	.2464	1679	.0975*	1141	.1977
Size						
ORG2a	3.545	.1699	1698	.0565*	1156	.1871
MNC/Subs conflict						
ORG2b	1.782	.4102	1656	.1439	1084	.6179
Subs/Subs conflict						
ORG3						
TP requirements						
ORG3a	5.440	.0659*	1546	.7843	1170	.1394
Ease/Cost						
ORG3b	1.409	.4944	1600	.3934	1032	.9318
Evaluation						
ORG3c	.293	.8638	1459	.5233	1023	.8500
Decision making						
ORG4						
Performance evaluation						
ORG4a	1.184	.5531	1543	.7759	1093	.4744
Subs.—nonincome						
ORG4b	.789	.6741	1531	.9112	982	.5159
Profit						
ORG4c	6.867	.0323**	1340	.0119**	885	.0214**
Subs.—income						
ORG4d	1.999	.3681	1655	.1606	1094	.5529
Innovation						
ORG5	.408	.8154	1513	.9427	1004	.6762
Compensation						
ORG6	*No significance on any of the nine objectives*					
Profit orientation						
ORG7	4.655	.0976*	1419	.0734*	.1014	.6290
Decentralization						

Notes: * Significant at alpha = .10.
      ** Significant at alpha = .05.

criteria. Compensation criteria (ORG5) and the profit orientation of the MNC (ORG6) were not significant.

The degree of decentralization (ORG7) was significant in both the Mkt/Neg/Cost and Cost/Other analyses. Each decision was measured using the following scale: 1 = decision made by subsidiary manager, 2 = decision made jointly by subsidiary manager and upper management, and 3 = decision made by upper management. Negotiated MNCs were the most decentralized, with a mean of 1.75, followed by market (1.88) and cost (1.98) MNCs. These findings are expected, because negotiation between managers requires a decentralized

MNC in order to succeed. The specific choice of a transfer pricing method was more centralized than the average; however, negotiated MNCs were still less centralized (2.33) than market (2.50) and cost (2.61) MNCs.

Results—Hypothesis 2

It was also hypothesized that international (external) variables do not influence the method choice. As shown in Table 5, several variables are significant, causing rejection of the second hypothesis.

The size of the transfers (INTL1) was not significant, contrary to prior findings. Also, no significant industry effects (INTL2) were found.

The international tax and tariff requirements of a transfer pricing method (INTL3a) were significant in the three analyses. The majority of market MNCs (69%) felt that these requirements were important when choosing their transfer pricing methods, compared with 42% of negotiated and 44% of cost MNCs. Market price allows an MNC to have a justifiable price to meet IRC Section 482 requirements regarding taxes, and to present to customs officials when tariffs were involved.

The economic stability of the parent MNC (INTL4a) was significant in the Mkt/Neg/Cost analysis. Market and cost firms were more stable (77% and 73%, respectively) than negotiated MNCs (42%). Perhaps the instability of the economy leads MNCs to consider negotiation as a better method of dealing with an uncertain environment than market or cost-based methods.

The favorableness of the economy to the parent MNC (INTL5a) was a significant factor only in Mkt/Other. Seventy-seven percent of market MNCS rated their economy as favorable, compared with 50% of negotiated and 61% of cost MNCs. Given the economic stability findings (INTL4a), market MNCs operate in a considerably more auspicious economic environment than negotiated MNCs.

MNC practices (INTL6) were significant in Mkt/Neg/Cost and Cost/Other analyses. Results for INTL6 and INTL7 are presented in Table 6, and present some interesting findings. The majority of market and cost MNCs indicate that they use two sets of books, either alone or in combination with approximating market conditions or adjusting the budget. Larger MNCs used two sets of books, either alone or with one of the other three practices, while smaller MNCs disregarded transfer pricing effects or adjusted the budget. In contrast, negotiated MNCs either adjusted the budget or approximated market conditions instead of keeping dual books. These findings are contrary to prior studies which found that most MNCs account for transfer price effects via budget adjustments.[6]

Because 82% of the MNCs use at least one practice to counter the problematic effects of transfer pricing, why have they then chosen their particular method? MNCs may use the method dictated by IRC Section 482

Table 5.   Analyses by International Variables Affecting
Transfer Pricing Method Choice

	Mkt/Neg/Cost		Full Cost/Other		Market/Other	
Variables	K-W	Signif	Wilcoxon	Signif	Wilcoxon	Signif
INTL1	.727	.6953	1484	.7078	1062	.8092
Transfer size						
INTL2	.002	.9988	1520	.9999	1037	.9777
Industry						
INTL3	4.555	.1025	1631	.2153	1169	.1255
TP requirements						
INTL3a	4.900	.0864*	1695	.0733*	1241	.0287**
Tax/tariff						
INTL3b	.233	.8900	1545	.8036	1031	.9269
Competition						
INTL3c	1.878	.3910	1651	.1839	1150	.2379
Currency						
INTL3d	2.528	.2826	1651	.1840	1185	.1191
Economy						
INTL4						
Stability						
INTL4a	5.998	.0498**	1478	.6093	1121	.2977
Parent Stability						
INTL4b	3.118	.2103	1446	.3841	1024	.8453
Subsidiary Stability						
INTL5						
Favorableness						
INTL5a	3.309	.1912	1574	.5312	1179	.0848*
Parent Favorableness						
INTL5b	2.873	.2377	1428	.3138	1041	.9954
Subsidiary Favorableness						
INTL6	8.514	.0142**	1275	.0131**	971	.4576
MNC practices						
INTL7	22.244	.0001**	1069	.0001**	708	.0002**
Section 482 regulations						

Notes: * Significant at alpha = .10.
         ** Significant at alpha = .05.

regulations instead of one better suited to their specific circumstances. These practices are dependent on the transfer pricing method chosen, and not vice versa.

The optimal method given IRC Section 482 regulations (INTL7) was significant in the three analyses. As shown in Table 6, MNCs other than market MNCs seem to pick their transfer pricing method based on other than IRC Section 482 dicta. Market (or comparable uncontrolled price) is the preferred IRC Section 482 method; however, 39% of current market MNCs do not think that this method is optimal for their companies and would use resale or a fourth

SUSAN C. BORKOWSK

Table 6. MNC Practices (INTL6) and IRS Section 482 Methods (INTL7) by Transfer Pricing Method Actually Used

	Market	*Negotiated*	*Cost*	*Total*
INTL6: Practices Used				
Use two sets of books	7 (26.9%)	2 (16.7%)	7 (17.1%)	16 (20.2%)
Approximate market conditions	2 (7.7%)	3 (25.0%)	2 (4.9%)	7 (8.9%)
Disregard transfer pricing effects	2 (7.7%)	3 (25.0%)	9 (22.0%)	14 (17.7%)
Adjust budget for transfer pricing effects	2 (7.7%)	3 (25.0%)	1 (2.4%)	6 (7.6%)
Use two sets of books plus another practice	13 (50.0%)	1 (8.3%)	22 (53.6%)	36 (45.6%)
Totals	26 (32.9%)	12 (15.2%)	41 (51.9%)	79 (100.0%)
INTL7: Optimal Method given IRS Section 482				
Comparable uncontrolled price (market)	16 (61.5%)	7 (58.3%)	8 (19.5%)	31 (39.2%)
Resale	8 (30.8%)	2 (16.7%)	2 (4.9%)	12 (15.2%)
Cost-plus	0 (0.0%)	3 (25.0%)	26 (63.4%)	29 (36.7%)
Other (the fourth method)	2 (7.7%)	0 (0.0%)	5 (12.2%)	7 (8.9)
Totals	26 (32.9%)	12 (15.2%)	41 (51.9%)	79 (100.0%)

method. It could therefore be assumed that these MNCs have chosen market primarily due to IRC Section 482 and the risk of an audit.

CONCLUSIONS

International transfer pricing methods are chosen because of certain organizational and international characteristics of the MNC's operating environment. Tax and tariff concerns, the stability of the parent MNC's economy, the ease and cost of implementation, the use of subsidiary profit for performance evaluation, and the degree of the MNC's decentralization all contribute to the choice of a method. The majority of the MNCs use cost-based methods, contrary to IRC Section 482 requirements and current theoretical recommendations.

Regardless of the method chosen, most MNCs employ some practice to counter the negative effects of that method. Dual sets of books, either alone or in conjunction with another practice, is the prevalent approach of MNCs, accounting for 66% of the sample. The findings of this study highlight the need to reevaluate the objectives of international transfer pricing. While a given transfer pricing method satisfies some of these objectives, it does not satisfy others. This causes MNCs to institute remedial practices to counter the negative aspects of their chosen methods.

In addition, the IRS requirements in IRC Section 482 do not provide the MNCs with what the latter consider optimal methods; however, some MNCs do choose their method because of IRC Section 482 rather than their situation-specific needs. This finding, coupled with recent research on IRC Section 482 audits and the issuance of the IRS White Paper, indicate a need to reevaluate and update the existing tax regulations and requirements to better reflect the realities of MNC practice rather than theory.

In addition to reevaluating MNC transfer pricing objectives and IRS requirements, another area of future research involves comparing the transfer pricing methods, objectives, and performance evaluation criteria of companies which transfer goods to both domestic and international subsidiaries.

This study involved MNCs based in the United States. Findings should not be generalized to non-U.S. based MNCs because tax, tariff, and customs regulations, as well as political, economic, and currency considerations, will differ.

NOTES

1.   The contingency theory framework used in the domestic transfer pricing study (Borkowski 1990) is equally applicable to international transfers. A separate expanded survey specifically dealing with MNC transfer pricing issues was developed for the current study. This expanded questionnaire was sent to a new sample of companies with international subsidiaries, whose responses are reported and analyzed in this study.

2.   Henceforth, any references to transfer pricing assume international, rather than domestic, transfers.

3.   Abdel-khalik and Lusk (1974) and Borkowski (1990) provide a detailed discussion of general transfer pricing theory and studies involving domestic transfer pricing.

4.   See Tang (1979) for a detailed discussion of the earlier studies.

5.   Metal firms are in the metals/mining and steel industries. Manufacturing firms include automotive, electronics, machinery, instruments, office equipment/computers, and tire and rubber companies. Process firms encompass the chemical, fuel/oil, and food processing industries, while any remaining industries are classified as other.

6.   Several respondents who identified themselves in order to receive results of this study, or to offer additional information, were contacted regarding INTL6 to verify their interpretation of the survey question. Their responses indicate that they understood the approaches described, in particular "Using two sets of books, one for tax, finance and local purposes, and the other for management and control purposes."

REFERENCES

Al-Eryani, M., P. Alam, and S. Akhter. 1990. Transfer pricing determinants of U.S. multinationals. *Journal of International Business Studies* 21: 409-25.

Arpan, J. 1972. *International Intracorporate Pricing.* New York: Praeger Publishers.

Benke, R., and J. Edwards. 1980. Transfer pricing: Techniques and uses. *Management Accounting* (June):44-46.

Benvignati, A. 1985. An empirical investigation of international transfer pricing by U.S. manufacturing firms. Pp. 193-211 in *Multinationals and Transfer Pricing*, edited by A. Rugman and L. Eden. New York: St Martin's Press.

Birnberg, J., and M. Shields. 1984. The role of attention and memory in accounting decisions. *Accounting, Organizations and Society* 9(3/4):365-82.

Bisat, T. 1966. *An Evaluation of International Intercompany Transactions*. Washington, DC: American University.

Borkowski, S. 1990. Environmental and organizational factors affecting transfer pricing: A survey. *Journal of Management Accounting Research* (Fall): 78-99.

Brooke, M., and H. Remmers. 1972. *The Strategy of Multinational Enterprise*. New York: Elsevier Publishing.

Burns, J. 1980. How IRS applies the intercompany pricing rules of Section 482: A corporate survey. *The Journal of Taxation* 52 (May): 308-14.

Business International Corporation. 1965. *Solving International Pricing Problems*. New York: BIC.

_____. 1973. *Setting Intra-Corporate Pricing*. New York: BIC.

Coburn, D., J. Ellis, and D. Milano. 1981. Dilemmas in MNC transfer pricing. *Management Accounting* 63 (November):53-69.

Eccles, R. 1985. *The Transfer Pricing Problem*. Lexington, MA: D.C. Heath.

Greene, J., and M. Duerr. 1970. *Intercompany Transactions in the Multinational Firm*. New York: The Conference Board.

Halperin, R., and B. Srindhi. 1987. The effects of the U.S. income tax regulations' transfer pricing rules on allocative efficiency. *The Accounting Review* 62 (October):686-706.

Horngren, C., and G. Foster. 1987. *Cost Accounting: A Managerial Emphasis, 6th ed.* Englewood Cliffs, NJ: Prentice-Hall.

Knowles, L., and I. Mathur. 1985. International transfer pricing objectives. *Managerial Finance* 11 (Spring):12-16.

Milburn, A. 1977. *International Transfer Pricing in a Financial Accounting Context*. Champaign, IL: University of Illinois.

Price Waterhouse. 1984. *Transfer Pricing Practices of American Industry*. New York: Price Waterhouse.

Rushinek, A., and S. Rushinek. 1988. Multinational transfer-pricing factors: Tax, custom duties, antitrust/dumping legislation, inflation, interest, competition, profit/dividend, and financial reporting. *The International Journal of Accounting* 23 (Spring): 95-111.

Schindler, G. 1988. Intercorporate transfer pricing. *The Tax Adviser* (May): 378-84.

Schindler, G., and D. Henderson. 1985. Intercorporate transfer pricing: 1985 survey of Section 482 audits. *Tax Notes* 29 (December 16): 1171-79.

Schweikart, J. 1985. Contingency theory as a framework for research in international accounting. *International Journal of Accounting* (Fall):89-98.

_____. 1986. Manager perception of the relevance of managerial accounting information. *Accounting, Organizations and Society* 11:541-54.

Shulman, J. 1966. *Transfer Pricing in Multinational Business*. Boston, MA: Harvard University Press.

Tang, R. 1979. *Transfer Pricing in the United States and Japan*. New York: Praeger Publishers.

_____. 1982. *Multinational Transfer Pricing*. Toronto: Butterworths.

U.S. Department of the Treasury and Internal Revenue Service. 1988. *A Study on Intercompany Pricing (White Paper)*. Washington, DC: Commerce Clearing House.

Vancil, R. 1978. *Decentralization: Managerial Ambiguity by Design*. Homewood, IL: Dow Jones-Irwin.

Watson, D., and J. Baumler. 1975. Transfer pricing: A behavioral context. *The Accounting Review* (July):466-74.

Wu, F., and D. Sharp. 1979. An empirical study of transfer pricing practice. *International Journal of Accounting* 14 (Spring): 71-99.

Yunker, P. 1983. A survey study of subsidiary autonomy, performance evaluation and transfer pricing in multinational corporations. *Columbia Journal of World Business* 18 (Fall):51-62.

SOURCES OF RISK IN U.S.-BASED MULTINATIONAL COMPANIES:
A COMPARATIVE CROSS-SECTIONAL TIME SERIES ANALYSIS

Pochara Theerathorn, Carlos Alcerreca-Joaquin, and Philip H. Siegel

ABSTRACT

During the period of 1973-1987, U.S.-based multinational corporations had lower returns and higher systematic risk than those of their domestic counterparts. The level of international involvement, dividend policy, financial leverage, asset structure, firm size, and profitability are used to explain variability in a stock's systematic risk. The period when beta is measured can also play a crucial role in determining a firm's performance.

Advances in International Accounting,
Volume 5, pages 193-209.
ISBN: 1-55938-415-8

INTRODUCTION

There has been a long history of accounting research that has focused on the effects of new and currently available information on stock prices and stockholders' returns. Early research by Ball and Brown (1968) and Beaver (1968) suggests that the events that affect accounting earnings also have an impact on security prices. Recent research suggests that the relationship between earnings and prices is not perfect (Beaver, Lambert, and Ryan 1987), either because other factors, in addition to earnings, have an impact on prices, or because not all the change in earnings may be associated with the future dividend-paying ability of the firm. This position has also been adopted by Stice (1991) who demonstrates that market inefficiencies exist for small firms whose earnings are announced late in the *Wall Street Journal* in relationship to their 10K filing. Additional evidence on this position is presented in a study by Cornell and Landsman (1989) which indicates that fourth quarter earnings announcements provide more information to decision makers than interim reports. Brown (1989) recently commented that this type of research incorporates very limited views of the financial world. Most of this research has primarily explored the relationship between earnings behavior and changes in stock prices. Less research has been done on the impact that accounting information has on the risk of the firm.

The usual list of the determinants of risk includes (Foster 1986): financial leverage, operating leverage, unexpected earnings variability, and lines of business (Beaver, Kettler, and Scholes 1970; Hill and Stone 1980). A different research avenue is represented by the attempt to explore whether different corporate strategies have an impact on the risk and return of the firm. Such is the case of corporate international diversification.

A debate has existed on whether corporate international diversification exists in order to increase or maintain profits, or to reduce risk (Fatemi 1984). Relevant research questions include: (1) whether U.S.-based multinational corporations (MNCs) are riskier than firms operating solely in the United States given the additional risk of international operations (political risk, exchange rate risk, etc.); (2) whether international diversification reduces the market risk of the multinational operations to the same level as, or even lower level than, the market risk of domestic corporations; and (3) whether higher risk-adjusted rates of return are obtainable from these MNCs than from domestic corporations (DCs). This research is useful to help understand the extent to which MNCs may have different objectives in their international diversification programs.

This type of research is of utmost importance in light of the issuance of SFAS 52. SFAS 52 requires managers to exercise judgment in using temporal or current rate methods of currency translation. Because return on investment is a key measure of subsidiary performance, the measurement of foreign subsidiary assets becomes critically important to evaluate managerial performance, determine incentives, and allocate resources. Most market and event empirical studies have assumed that different accounting methods do not have an effect on the free cash flow of the firm, and hence no effect on the price of the common equity. However, it has been argued that the choice of accounting methods will affect the monitoring, information, and other contracting costs between the firm and its stakeholders and, consequently, the contracting parties' wealth (Watts and Zimmerman 1990).

The adoption of SFAS 52 was generally expected to reduce earnings volatility (Jaggi and Chhatwal 1990). Several recent studies have empirically studied the relationship between the adoption of SFAS 52 reporting standards and earnings volatility. Kirsch, Evans, and Doupnik (1990) suggest that firms differ in the extent to which they emphasize short-term accounting objectives over economic value objectives. Their research found that firm size and absence of a management compensation scheme were the determining factors in whether firms supported the adoption of SFAS 52 standards. Similarly, Ndubizu (1990) found that early adopters of SFAS 52 were smaller in size, had lower earnings growth, higher dividend payouts, and lower earnings volatility than later adopters. The study suggests that reductions in earnings volatility were the principal reasons for corporations reporting under SFAS 52 standards. Finally, a study by Shalchi and Hosseini (1990) explored the ability of SFAS 52 to reduce the amount of unnecessary hedging activities of MNCs to reduce earnings volatility which resulted from exchange rate fluctuations. The study found that the hedging activities were decreased by MNCs after the adoption of SFAS 52.

In light of this past research on accounting data studies, it is important to consider what are the sources of the market risk of MNCs and DCs. These past studies have suggested that research is needed to analyze what are the specific sources of risk in order to develop a more effective methods to evaluate foreign subsidiary performance.

PREVIOUS EMPIRICAL RESEARCH

It is useful to review the literature that focuses on the questions addressed in the prior section. Hughes, Logue, and Sweeny (1975) addressed the question of whether MNCs are a substitute for direct international diversification. The study compared the performance of a portfolio of 46 MNCs to a portfolio of 30 DCs in the period 1970-1973. The study concluded that MNCs had lower systematic risk, lower total risk, and higher risk adjusted returns than DCs.

The question of risk reduction through international diversification was analyzed by Rugman (1976). The study analyzed the variance of the accounting rate of return on equity of a sample of Fortune 500 firms in the period 1960-

1969. The study found that the variance decreased as the degree of international involvement increased. The study predicted that the benefits of indirect international diversification through the purchase of MNC stock would disappear as the world economy became more integrated.

The same question was addressed by Agmon and Lessard (1977), who studied 217 firms in the period 1959-1972, using an international market model. The study found that when the degree of international involvement of the firm is high, (a) the beta coefficient relating firm return to the U.S. market index is low, and (b) a rest-of-the-world beta is high. This study lends additional support to the argument that indirect international diversification is an appropriate method of portfolio risk reduction.

Not all studies agree with this position. For example, Jacquillat and Solnik (1978) analyzed 23 U.S. MNCs and 40 European MNCs during the period 1966-1974, and concluded that MNC's stock prices behaved so closely to the prices of domestic firms that investing in MNCs is not a substitute for direct international diversification. Similarly, Senchack and Beedles (1980) found that portfolios consisting of DCs provided lower levels of total risk with fewer securities than the corresponding portfolios of MNCs.

Additional support for the position than MNCs are imperfect substitutes for direct international diversification is presented by Brewer (1981). Employing a Fama and MacBeth (1973) methodology, Brewer compared the risk adjusted return of 137 DCs and 151 MNCs for the period 1963-1975. The study found that the two groups of firms had similar risk-adjusted levels of performance. Therefore, investing in MNCs does not provide any special advantages over investing in DCs. However, the study does recognize that investing in MNCs may reduce the beta of the portfolio of the U.S. investor.

The question of the effect of international operations on the market value of the firm was also considered by Errunza and Senbet (1981), who approached this question by using a sample of 50 firms for which they calculated the ratio of excess market value to sales. The analysis found a positive correlation between this ratio and the degree of international involvement. This positive association was particularly significant for the time that barriers to international capital flows were in effect. The results of the study suggest that both real and capital market imperfections have an effect on the rate of return of MNCs.

Fatemi (1984) further considered the shareholders' benefits from corporate international diversification. He compared the risk and return of 84 MNCs and 52 DCs in the period 1976-1980. The study found that the riskiness of the MNCs declines as the degree of international involvement increases. Using the Fama and MacBeth methodology, the risk-adjusted returns were found to be identical for both groups. However, Fatemi found evidence of abnormal returns during the time period when the initial international diversification took place. This result highlights the significance of the time dimension factor in comparative studies.

Finally, Lee and Kwok (1988) compared 421 DCs and 413 MNCs during the period 1964-1983. The study analyzed the foreign tax ratio, instead of the traditional foreign sales ratio, in order to classify firms as to whether they were DCs or MNCs. The results of the study were that MNCs were less leveraged than DCs; this was explained by higher agency costs that decreased the optimal level of MNCs' debt.

The recent empirical research on the risk and return of MNCs suggests several research avenues:

1. obtaining larger sample sizes by using the foreign tax ratio to determine the degree of international involvement;
2. comparing several time periods to study the relationships between changes in the firm's degree of international involvement and changes in performance, to determine whether increases in the degree of international involvement generate increases in stockholder return or decreases in the level of risk of the firm; and
3. identifying the sources of the level of risk of MNCs to suggest key performance indicators to measure and monitor as part of the corporation's control system.

This paper will build on these recommendations and will proceed in the following manner: We first describe the data, sample selection, and methodology used to segregate MNCs and DCs, and report results of hypotheses tests. The next section analyzes the risk measure in order to identify the sources of variation in the sample companies' systematic risk. Finally, we summarize our findings and recommend areas for additional research.

DATA AND METHODOLOGY

The sample consists of NYSE-listed companies that have continuous data—both monthly market and annual financial data—in the COMPUSTAT tapes during any of the three periods: 1973-1977, 1978-1982, and 1983-1987. We omitted those with SIC code 4000-4999 to avoid regulated firms. The purpose is to include as many companies as possible so that a very general picture would emerge as to the risk-return structure among companies with varying degrees of international involvement. To require continuous data for the entire 15 years might leave out a substantial number of firms that could contribute to a fuller picture. Collected this way, the sample offers a cross-section of all companies: long-established as well as recent ones. Accordingly, a firm may be counted as one, two, or three cases, depending on whether it has continuous financial and market data in one or more of the 5-year periods. Altogether, we obtained 899 cases for this study.

Next, we calculated the degree of international involvement (DII) for each firm. DII is defined as the ratio of foreign taxes to total taxes, averaged over the 5-year period. Foreign taxes are a better gauge of a company's foreign involvement in international investment than, say, annual sales or the number of subsidiaries overseas. In the first case, using overseas sales as a proxy for DII mixes international investment with international trade; in the second, the number of subsidiaries is a rather qualitative measure: subsidiaries are not identical, nor even homogeneous, and therefore not comparable even though they belong to the same parent company. Foreign taxes, on the other hand, accurately represent the level of a company's profit-making activities in a host country and can, therefore, approximate the extent to which it is involved business-wise in that country. Besides, data on foreign taxes have been reported in financial statements since 1969, and are more easily obtainable than the segregated amount of overseas sales or number of overseas subsidiaries.

The companies were divided into three groups according to their DII: domestics (0-10%), intermediate (10%-30%), and multinationals (30% and over). Table 1 reports the breakdown of sample companies by group as well as by time period. Panel A indicates that there are 336 DCs, 276 ICs, and 287 MNCs. Panel C shows that the average DC has a DII of about 2.2%; the average IC, 19.4%; and the average MNC, 49.5%.

Each company's monthly returns, systematic risk, and excess returns were then calculated, where

r_{it} = unadjusted return for firm i in month t,

   = $P_{it}/P_{i,t-1} - 1$; P_{it} is the price per share adjusted for stock split and stock dividend.

B_{it} = beta of company i calculated by using returns of the 60 months prior to month t,

   = $\text{cov}(r_i, r_m)/\text{var}(r_m)$, and r_m is the return from a market portfolio proxied by the S&P's 500 index.

e_{it} = excess return of firm i in month t,

   = $r_{it} - Y_{0t} - Y_{1t}B_{it}$; Y_{0t} and Y_{1t} are market-determined regression coefficients of returns on systematic risk calculated according to the methodology of Fama and MacBeth (1973).

Next, we calculated average monthly returns, betas, and excess returns of the portfolios integrated by all the domestic, intermediate, and multinational companies. A portfolio average is simply the arithmetic average of all values in that portfolio. In all, we obtained 180 monthly returns, betas, and excess returns for each of the three portfolios over January 1973-December 1987.

Table 1. The Sample

A. Distribution of Companies

Degree of International Involvement

Time Period	0-10%	10-29%	29%+	Total
1973-1977	57	67	57	181
1978-1982	142	126	110	378
1983-1987	137	83	120	340
Total	336	276	287	899
	DCs	ICs	MNCs	

B. Distribution in Percentage

Degree of International Involvement

Time Period	0-10%	10-29%	29%+	Total
1973-1977	6.34	7.45	6.34	20.13
1978-1982	15.79	14.02	12.24	42.05
1983-1987	15.24	9.23	13.35	37.82
Total	37.37	30.70	31.93	100.00

C. Average DII in Each Subgroup

Degree of International Involvement

Time Period	0-10%	10-29%	29%+	Total
1973-1977	2.86	19.50	48.14	23.28
1978-1982	1.89	19.51	47.72	21.10
1983-1987	2.30	19.11	51.76	23.86
15-year average	2.22	19.38	49.49	22.58

Statistical Analysis

Table 2 contains the summary statistics for the distributions of portfolio returns, betas, and excess returns. On average, MNCs as a group provide the lowest return, with DCs the highest, and ICs in between. The portfolio betas, however, are not in the same order: ICs' average is the lowest, MNCs' the highest, and DCs' in between and somewhat closer to MNCs than to ICs. Similar to Fatemi's (1984) results, the average excess return of DCs is highest, followed by that of ICs, and with MNCs' average being negative. For MNCs, the standard deviation of excess returns and the coefficient of variation of monthly and excess returns are also the highest, followed by those of ICs and DCs, respectively.

The objective here is to test whether the three portfolios are statistically equal with respect to monthly returns, betas, and excess returns. To determine the appropriate statistical technique for the test, we need to find out whether the data are normally distributed. If so, we can use a *t*-test to compare the means

Table 2. Monthly Returns, Betas, and Excess Returns
Summary Statistics (180 months)

	Monthly Portfolio Returns	Portfolio Betas	Excess Returns
A. Domestic Companies			
Mean	1.128	1.177	.243
Standard Deviation	6.375	.154	1.202
Minimum Value	−29.511	.948	−2.721
Maximum Value	23.885	1.512	4.402
Std Error of Mean	.475	.011	.090
Coeff of Variation	5.652	.131	4.947
Kolmogorov-Smirnov Z	.942	1.551*	.580
Significance of Z	.337	.016	.889
B. Intermediate Companies			
Mean	1.020	1.140	.181
Standard Deviation	6.170	.095	1.233
Minimum Value	−29.069	.995	−4.441
Maximum Value	18.103	1.373	4.606
Std Error of Mean	.460	.007	.092
Coeff of Variation	6.049	.083	6.812
Kolmogorov-Smirnov Z	.931	1.812**	.969
Significance of Z	.351	.003	.305
C. Multinational Companies			
Mean	.704	1.197	−.159
Standard Deviation	6.537	.102	1.285
Minimum Value	−30.207	1.041	−3.320
Maximum Value	23.561	1.399	3.116
Std Error of Mean	.487	.008	.096
Coeff of Variation	9.286	.085	−8.082
Kolmogorov-Smirnov Z	.833	1.739**	.539
Significance of Z	.491	.005	.933

Notes: * significant beyond the .05 level.
** significant beyond the .01 level.

of each sample-pair; otherwise, the Kruskal-Wallis one-way analysis of variance is the proper test for equality. For the normality test, the Kolmogorov-Smirnov statistics were computed for actual returns, betas, and excess returns of all three groups. According to these statistics, also listed in Table 2, all actual returns and excess returns follow a normal distribution, whereas betas do not.

After the appropriate statistical techniques were applied, the results presented in Table 3 indicate that the monthly returns, excess returns, and betas from DC and MNC portfolios are not samples from the same population. In all three aspects, MNCs as a group are significantly different from ICs and DCs, which are not statistically distinguishable between themselves. These

Table 3. Test of Equality of Portfolio Returns, Betas,
and Excess Returns

I.   T-Test (179 d.f.)

A. Monthly Portfolio Returns

	X_1	X_2	X_1-X_2	std dev	std err	t-value
DCs : ICs	1.1282	1.0200	.1083	1.818	.136	.80
DCs : MNCs	1.1282	.7045	.4238	1.725	.129	3.30**
ICs : MNCs	1.0200	.7045	.3155	1.429	.106	2.96**

B. Excess Portfolio Returns

DCs : ICs	.2435	.1810	.0625	1.771	.132	.47
DCs : MNCs	.2435	−.1586	.4021	1.648	.123	3.27**
ICs : MNCs	.1810	−.1586	.3396	1.362	.102	3.35**

II. Kruskal-Wallis One-Way Analysis of Variance of Ranks of Betas

	Mean Rank				Observed Level of Sig.
	DCs	ICs	MNCs	H-stat	
DCs : ICs : MNCs	263.05	235.90	312.56	22.3406	.0000**
DCs : ICs	86.76	74.24		1.3008	.2541
ICs : MNCs		152.15	208.85	26.7113	.0000**
DCs : MNCs	94.21		166.79	6.2466	.0124*

Notes: * significant beyond the .05 level.
   ** significant beyond the .01 level.

results are somewhat different from those reported in previous studies. Fatemi (1984) and Brewer (1981), for example, found that the monthly returns from a portfolio of MNCs are not statistically different from returns from a portfolio of DCs, using 25% degree of international involvement as the cutoff point for MNCs. But our results confirm those of Fatemi's in terms of excess returns and betas, although our level of significance is not as high.

The MNC portfolio's low monthly and excess returns, combined with its high beta relative to the other portfolios, imply that MNC stock was overpriced. Investors may have perceived MNCs to be less risky than DCs of comparable size and industry while in actuality they were not. The realized returns then fell short of the expected returns, causing the excess returns to be negative. Perhaps the market imperfections thought to exist during the 1960s have lessened. More recently, a larger number of MNCs has offered the same diversification service, thereby reducing the "diversification premium" predicted by Errunza and Senbet (1981). Our findings confirm the results of the studies by Jacquillat and Solnik (1978), Senchack and Beedles (1980), and Brewer (1981). These results are also consistent with the idea that during the

1960s, foreign economies were attractive enough to justify an investment in projects with a positive net present value; the price of MNC stock was adjusted upward, thus producing a high rate of return during the period when the new investments were made. In the 1970s and in the 1980s, the number of extraordinary investment opportunities decreased, however, the perception of lower-risk MNCs may have persisted. The investors accordingly overpriced MNC stock, receiving negative risk-adjusted returns as a result.

DETERMINANTS OF THE SYSTEMATIC RISK

There is still speculation about the sources of risk in MNCs. Agmon and Lessard (1977) concluded in their study that a firm's degree of international involvement is negatively correlated with its beta coefficient. Yet, Jacquillat and Solnik (1978), and later Senchack and Beedles (1980), found that DII is not a very important determinant of a company's systematic risk. This controversy gave rise to the speculation about other factors which might influence a company's beta.

Time period under study has been cited as a possible explanatory variable. One supporting hypothesis is that the global integration of product and factor markets has resulted in MNCs having the same investment opportunity schedule as DCs, thereby reducing the former's advantage in international diversification. Errunza and Senbet (1981), and later Siddharthan and Lall (1982), found time period an important factor affecting the rate of growth and degree of international involvement of MNCs.

Financial and business risk have also proven to be significant factors in the majority of studies. Hill and Stone (1980) found the market betas dependent on both financial structure and systematic operating risk (defined as the covariance between company and market return on assets divided by the variance of market return on assets). However, Shaked (1986) and Lee and Kwok (1988) found that MNCs are less leveraged than DCs. Industry, company size, and fixed assets have also been used as measures of business risk to explain the variation in market betas. The agency costs of debt, proxied by advertising and R&D expenses, seem to play a significant part in influencing a firm's capital structure—and hence its beta (Myers 1977). Finally, profitability and dividend policy can also affect betas: a company that maintains a stable level of earnings, or has been paying stable dividends, or both, represents a lower risk to investors than a less predictable company.

The purpose of this part of our study is to identify factors that have significant influence on the systematic risk of a company, multinational or domestic, and to determine whether there are different sets of factors for MNCs and DCs, or whether common factors exist that affect all companies. So we calculated, for each of the 899 cases, a 5-year average of the following variables:

1. EPS = earnings per share
2. ADV = advertising expenses as a percentage of sales
3. RD = R&D expenses as a percentage of sales
4. PROFITS = net profit margin
5. FIXASSET = ratio of net fixed assets to total assets
6. DEBT = debt/equity ratio
7. DVD = dividend payout ratio
8. DYLD = dividend yield
9. SALE = total sales

Analysis of Covariance

The nine variables above, together with time period, industry factor (1-digit SIC code), and the degree of international involvement were used as predictor variables for betas. Because some of the variables are metric, while others are categorical variables, an analysis of covariance was performed, with beta as the criterion variable, time period, industry and DII as factors, and the remaining variables as covariates.

The ANCOVA results are presented in Table 4, Panel A. Time period, DII, and industry effects are all statistically significant sources of variation in the systematic risk. As for the covariates, dividend yield, advertising expenses, dividend payout ratio, and annual level of sales are significant. We then repeated the analysis, this time with DII as a metric variable, and obtained a significant coefficient (see Table 4, Panel B) for DII and the dividend payout ratio in addition to DYLD and ADV.

To identify the effects of these explanatory variables on the systematic risk of companies with varying degree of international involvement, the ANCOVA was next performed on each of the three subgroups. The results, in Table 5, indicate that DII is significant neither as a factor nor a covariate in any of the three portfolios. For the primary factors, both time period and industry effects are significant for ICS, but only time period is significant for DCs, and neither is significant for MNCs. As for the covariates, only dividend yield is a common significant covariate for all three groups. The betas of DCs and MNCs are significantly influenced by advertising expenses and business risk as proxied by the level of fixed assets. However, the negative sign of the FIXASSET coefficient was unexpected: firms with high operating leverage were expected to have a higher beta. This result suggests that firms tend to invest a large fraction of their resources in fixed assets when the overall beta risk of the firm is below average. Another interesting result is that MNCs and ICs have net profit margin as a common significant covariate. This implies that, the more a firm is engaged in international business, the lower net profit margin it can garner.

Table 4.   Analysis of Covariance—Total Sample

A. Factors: Time Period, Industry, and Degree of International Involvement

	F-value	Sig of F
Regression	17.33	.000**
Time period effect	103.66	.000**
Industry effect	5.57	.000**
DII	5.11	.006**

Significant Covariates	Coefficient	t-value
DYLD	−.051	−9.05**
ADV	−.013	−3.54**
DVD	.003	2.75**
SALE	−.019	−2.26*

B. Factors: Time Period and Industry

	F-value	Sig of F
Regression	15.16	.000**
Time period effect	101.09	.000**
Industry effect	5.93	.000**

Significant Covariates	Coefficient	t-value
DYLD	−.051	−9.02**
ADV	−.013	−3.54**
DVD	.003	2.67**
DII	.143	2.69**

Notes: * significant beyond the .05 level.
          ** significant beyond the .01 level.

The negative signs of ADV, the ratio of advertising expenses to sales, may be an indicator of the extent of the "underinvestment problems" of the firm (Myers 1977). High advertising expenses indicate that an important part of the value of the firm is represented by investment options or intangible assets. Myers argues that firms with many intangible assets bear high agency costs in the form of discounts in the price of their bonds and a higher cost of debt: a situation that results in a lower optimal amount of debt in their capital structure. That is, high advertising expenses may be associated with a lower debt ratio and a lower beta. Perhaps there is another interpretation for ADV. Suppose we consider the ratio of advertising expenses to sales as a measure of investors' familiarity with a company. Fluctuations in the financial markets may not affect the price of a "household name" as much as that of a less familiar one, so its beta would be negatively correlated with the "degree of consumer familiarity" represented by the amount of advertising expenses as a fraction of sales.

Table 5. Analysis of Covariance—International
Involvement Subsamples

A. DCs

	F-value	Sig of F
Regression	6.97	.000**
Time period effect	42.88	.000**
Industry effect	1.78	.133

Significant Covariates	Coefficient	t-value
DYLD	−.053	−5.54**
DVD	.003	2.44**
DEBT	.002	2.33*
ADV	−.016	−1.97*

B. ICs

	F-value	Sig of F
Regression	6.81	.000**
Time period	27.04	.000**
Industry	4.21	.003**

Significant Covariates	Coefficient	t-value
DYLD	−.058	−5.36**
PROFITS	−.009	−2.36*
SALE	−.038	−2.28*

C. MNCs

	F-value	Sig of F
Regression	8.87	.000**
Time period	41.98	.000**
Industry	1.80	.129

Significant Covariates	Coefficient	t-value
DYLD	−.045	−4.69**
PROFITS	−.012	−3.56**
ADV	−.014	−2.82**
FIXASSET	−.003	−2.14*

Notes: * significant beyond the .05 level.
** significant beyond the .01 level.

To obtain further insight into the characteristics of systematic risk, the betas were partitioned by time period and DII. Table 6 lists the average beta of each subgroup as well as the column and row averages. The column totals are labeled "weighted averages" because the number of firms in each of the three time

Table 6.   Average Beta by Time Period and Degree of
International Involvement

| | Degree of International Involvement | | | 5-year average |
	under .10	.10-.299	.30 and above	
1973-1977	1.354	1.240	1.399	1.326
1978-1982	1.033	1.069	1.085	1.060
1983-87	1.012	1.036	1.051	1.032
Weighted average	1.079	1.101	1.133	1.103
	DCs	ICs	MNCs	

periods differs (see Table 1). From Table 6, we can see that MNCs' betas have always been higher than those of DCs. In particular, during 1978-1982 and 1983-1987, betas for DCs, ICs, and MNCs are nicely ranked in the same order as the DII. Note that betas for all groups decrease as we move from the earlier to the more recent period and become almost equal in the last one, a result that may be related to the integration of the world economy.

Multiple Regression Analysis

As some of the significant covariates seem to represent the same characteristic, the problem of multicollinearity may have existed in our ANCOVA results. So we applied a multiple regression analysis, using the stepwise method, on the entire sample as well as on each subgroup, with betas as dependent variables and all the metric variables (including DII) as predictors. The regression results appear in Table 7. These results are quite similar to those obtained by ANCOVA. For the entire sample, DII is significant and positive, implying that a firm's systematic risk increases with an increase in international involvement. Profitability, in the form of EPS, is also a significant predictor variable with the correct negative sign. FIXASSET still maintains a negative sign. The dividend policy variable (DYLD) is significant everywhere, perhaps an indication that firms with low beta risk can afford to pay a higher dividend.

DCs' systematic risk seems to be more significantly affected by the fixed assets ratio (FIXASSET) and the debt-to-equity ratio (DEBT) whereas the systematic risk of ICs and MNCs is not. Apart from the common factor of dividend yield (DYLD), MNCs and DCs appear as two entirely different groups of firms: MNCs' betas are influenced by firm size (SALE) whereas DCs' betas are affected by the advertising ratio (ADV). PROFITS, the only other common factor between them, even has opposite signs. A plausible interpretation of this result is that, in our sample, the more profitable MNCs have more diversified sources of corporate income whereas the more profitable DCs belong to cyclical industries and are subject to the ups and downs of the

Table 7. Multiple Regression Analysis Results
(t-value in parentheses)

	Total Sample	DCs	ICs	MNCs
Constant	1.342	1.320	1.510	1.546
	(37.45)	(24.24)	(14.35)	(16.68)
DII	.107			
	(2.03)			
EPS	−.018			
	(-3.31)			
ADV	−.018	−.027	−.021	
	(-4.66)	(-3.02)	(-3.28)	
PROFIT		.006		−.007
		(2.87)		(-2.42)
FIXASSET	−.003	−.004		
	(-3.34)	(-3.11)		
DEBT		.002		
		(2.06)		
DYLD	−.033	−.032	−.041	−.030
	(-6.06)	(-3.37)	(-4.13)	(-3.07)
SALE			−.032	−.040
			(-2.51)	(-2.77)
R^2	.10	.10	.12	.13

domestic business cycle. As for the ICs, their betas are affected by dividend policy, consumer familiarity (like the DCs), as well as by firm size (like the MNCs).

SUMMARY AND CONCLUSION

We examined the monthly return data and annual financial data of a sample of NYSE-listed companies over three 5-year periods, dividing those companies into three groups according to their degree of international involvement. We found that the returns and systematic risk of MNCs are significantly different from those of DCs, and that MNCs as a group have significantly higher systematic risk while registering significantly lower actual and risk-adjusted returns than DCs. It is possible that DC's lower values and higher returns are taking into account the threat of new foreign entrants into the domestic market, a risk that is not reflected in the current estimate of systematic risk. Upon further investigation, we discovered that the time period when data was collected for study and the industry to which a company belongs have, to a certain extent, a more significant bearing on the company's systematic risk than its degree of international involvement; that the betas of DCs and MNCs are similarly influenced by the corporate dividend policy, but DCs' betas are significantly affected by the amount of financial leverage and their asset structure, whereas MNCs' betas are influenced by company size and profitability.

Finally, it appears that the capital markets of the world are well integrated, so that the return of MNCs' stock is highly correlated with a domestic market index—as is evident in MNCs' high betas. The investors have certainly recognized investing in MNCs as a substitute for direct investments in foreign corporation; in fact, from our results they have placed too high a value on that substitutability.

Several recommendations for further research seem relevant. Because previous research found that earnings volatility did not have a clear impact on the adoption of SFAS 52, it is important to study whether market risk has been affected by the adoption of this accounting method. The question of whether emphasis on accounting or economic objectives have an impact on the type of managerial compensation schemes could be studied taking into account both risk and return considerations. The use of international segment information could be used to further validate the use of the foreign tax ratio as a measure of the degree of international involvement of the firm.

REFERENCES

Agmon, T., and D.R. Lessard. 1977. Investor recognition of corporate international diversification. *Journal of Finance* 32(September): 1049-55.

Ball, R., and P. Brown. 1968. An empirical evaluation of accounting income numbers. *Journal of Accounting Research* (Autumn): 159-78.

Beaver, W.H. 1968. The information content of annual earnings announcements. *Journal of Accounting Research* (Supplement): 67-92.

Beaver, W.H., P. Kettler, and M. Scholes. 1970. The association between market-determined and accounting-determined risk measures. *The Accounting Review* (October): 654-82.

Beaver, W.H., R. Lambert, and S. Ryan. 1987. The information content of security prices: A second look. *Journal of Accounting and Economics* (July): 139-57.

Brewer, H.L. 1981. Investor benefits from corporate international diversification. *Journal of Financial and Quantitative Analysis* 16(March): 113-26.

Brown, P. 1989. Ball and Brown. *Journal of Accounting Research* (Supplement): 202-17.

Cornell, B., and W.R. Landsman. 1989. Security price response to quarterly earnings announcements and analyst's forecast revisions. *The Accounting Review* (October): 680-92.

Errunza, V.R., and L.W. Senbet. 1981. The effects of international operations on the market value of the firm: Theory and evidence. *Journal of Finance* 36(2): 401-17.

Fama, E.F., and J.D. MacBeth. 1973. Risk, return, and equilibrium: Empirical tests. *Journal of Political Economy* (June): 607-36.

Fatemi, A.M. 1984. Shareholder benefits from corporate international diversification. *Journal of Finance* 39(5): 1325-44.

Financial Accounting Standards Board. 1981. *Statement of Financial Accounting Standards No. 52*, Foreign Currency Translation. Stamford, CT: FASB.

Foster, G. 1986. *Financial Statement Analysis*. Englewood Cliffs, NJ: Prentice-Hall.

Hill, N.C., and B.K. Stone. 1980. Accounting betas, systematic operating risk, and financial leverage: A risk-composition approach to the determinants of systematic risk. *Journal of Financial and Quantitative Analysis* 15(September): 595-637.

Hughes, J.S., D.E. Logue, and R.J. Sweeney. 1975. Corporate international diversification and market assigned measures of risk and diversification. *Journal of Financial and Quantitative Analysis* 10(4): 627-37627-37.

Jacquillat, B., and B.H. Solnik. 1978. Multinationals are poor tools for diversification. *Journal of Portfolio Management* 4(Winter): 8-12.

Jaggi, B., and G. Chhatwal. 1990. Impact of SFAS 52 on the accuracy of analysts' earnings forecasts. *Advances in International Accounting* 3: 139-54.

Kirsch, R.J., T.G. Evans, and T.S. Doupnik. 1990. FASB Statement 52, an accounting policy intervention: U.S.-based multinational corporate preenactment lobbying behavior. *Advances in International Accounting* 3: 155-72.

Lee, K.C., and C.C.Y. Kwok. 1988. Multinational corporations vs. domestic corporations: International environmental factors and determinants of capital structure. *Journal of International Business Studies* 19(2): 195-217.

Myers, S.C. 1977. Determinants of corporate borrowing. *Journal of Financial Economics* (November): 147-75.

Ndubizu, G.A. 1990. Earnings volatility and the corporate adoption decision on FASB Statement No. 52: An empirical analysis. *Advances in International Accounting* 3: 173-88.

Rugman, A. 1976. Risk reduction by international diversification. *Journal of International Business Studies* 7(Fall/Winter): 75-80.

Senchack, A.J., and W.L. Beedles. 1980. Is indirect international diversification desirable? *Journal of Portfolio Management* 6(Winter): 49-57.

Shaked, I. 1986. Are multinational corporations safer? *Journal of International Business Studies* 17(1): 83-106.

Shalchi, H., and A. Hosseini. 1990. The impact of FASB Statement No. 52 on foreign exchange contracts: An empirical evaluation. *Advances in International Accounting* 3: 189-201.

Siddharthan, N.S., and S. Lall. 1982. The recent growth of the largest U.S. multinationals. *Oxford Bulletin of Economics and Statistics* 44(1): 1-13.

Stice, E. 1991. The market reaction to 10-K and 10-Q filings and to subsequent the *Wall Street Journal* earnings announcements. *The Accounting Review* (January): 42-55.

Watts, R.L., and J.L. Zimmerman. 1990. Positive accounting theory: A ten year perspective. *The Accounting Review* (January): 131-56.

PART IV

CROSS-CULTURAL STUDIES

INTERNATIONAL AUDITING:

CROSS-CULTURAL IMPACTS ON QUALITY

James C. Lampe and Steve G. Sutton

ABSTRACT

The audit profession has been faced with numerous challenges in recent years. Two conditions that will likely continue to increase in the foreseeable future are a demand for increased audit quality and the complexity of audits performed in an international environment. This study identifies and examines the factors influencing audit quality and how these factors vary in an international setting. Specifically, the study first examines cross-cultural influences in the business environment as viewed by prior management research. Nominal group techniques are subsequently applied to involve auditors with experience in performing audits in both U.S. and non-U.S. locations in discussions concerning the variables impacting audit quality. Within these discussions, a primary objective was to identify the major dimensions impacting achievement of desired quality levels in international settings. Three dimensions were identified as capturing the primary cross-cultural influences on audit quality: (1) English versus non-English speaking, (2) Western versus Eastern culture, and (3) developed versus third world nations. Additionally, 17 audit quality factors were identified as having a critical

Advances in International Accounting,
Volume 5, pages 213-236.
Copyright © 1992 by JAI Press Inc.
All rights of reproduction in any form reserved.
ISBN: 1-55938-415-8

affect on domestic audits, and another 14 factors were identified as being critical
to international audits as one or more of the cross-cultural dimensions differed
for a given audit site.

The Securities and Exchange Commission recently reported that world
financial market activity approximately tripled from a level of $1,700 billion
in 1978 to over $5,000 billion in 1986 (SEC 1987). Not surprisingly, the
Japanese share of trade increased during this time period while the U.S. share
decreased from 52 to 43%. Paralleling the increased activity in international
financial markets, multinational corporations (MNCs) also reflected the
globalization of business during the 1980s. Once again, although losing in share
to the Japanese, U.S.-based MNCs experienced much more growth than purely
domestic counterparts. The rapid growth in international financial market and
MNC activity has also caused increased international audit activity by U.S.-
based CPA firms, by U.S.-based MNC internal audit departments, and by non-
U.S. auditors. One obvious evidence of rapidly increasing international activity
by U.S.-based CPA firms is the merger activity of the large firms. Reports
of the Institute of Internal Auditors also provide evidence that the decade of
the 1980s brought about increased education, training, and utilization of U.S.-
based MNC internal auditors (Vinton 1991). Non-U.S. auditors performing
fieldwork in the United States is also on the rise, such as when auditors from
French, Japanese, and Italian manufacturing firms performed pre-acquisition
audits on Firestone Rubber Company.

The common thread to all of the increasing international audit activity
described in the preceding paragraph is utilization of the work and reports
from an audit in one country by investors, managers, other auditors, and/or
other report users in different countries. A primary objective of an audit is
to improve service quality in order to make the audit more effective and
efficient. Greater effectiveness, at equal or lower cost, improves audit quality
by providing higher levels of assurance to the readers of the audit report.
Improved efficiency is attained when an equal or higher level of assurance is
provided and lesser time or other resources are used in the audit process.
Comprehensive and objective indicators of audit quality are difficult in purely
domestic audit engagements and many of the problems are magnified in
international audits. Similarly, audit practitioners and researchers alike have
noted that numerous problems and obstacles exist in both the performance
and research of international audits which are not present in domestic audits
(Needles 1989; Hussein, Bavishi and Gangolly 1986; Stamp and Moonitz 1982a,
1982b). Commonly referenced obstacles include: (1) lack of harmonization of
international audit standards, (2) diversity of language and communication
problems, and (3) cultural differences. Research by the International Section

of the American Accounting Association ranked these problems among those most in need of further research in order to provide improved understanding of international auditing and a conceptual basis for future development (Scott 1980; Choi 1981).

PRIOR RESEARCH IN INTERNATIONAL AUDITING

Through the decade of the 1980s and into the 1990s, most published research in international auditing has been primarily descriptive or comparative in approach and mostly oriented to harmonization of the standards applied during engagements or the report issued at the end of international engagements. A Canadian Government Accountants research monograph (Most 1988) provides an excellent summary and reference for numerous studies of international standards and reporting problems. Several different models have been proposed to classify international accounting and auditing research (Mueller 1979; Wallace 1987; Bindon and Gernon 1987). All three of these models recognize the dominance of descriptive and comparative studies in recent research. One reviewer in commenting on the descriptive emphasis in research notes that:

> commentators underestimate the problem when they complain of too much descriptive work. The real problem in international accounting is much worse: too much *inaccurate* descriptive work (Nobes 1983).

This apparent research emphasis on description and comparison is not due to the lack of other problems to investigate. Numerous publications have indicated a wide range of other problems facing international accounting and auditing (Aranya and Armenic 1981; Buckley and O'Sullivan 1980; Dykxhoorn and Skinning 1981; Gul and Yap 1984; Lee 1978; Mann and Redmayne 1979; Ricchiute 1978; Tantuico 1980; Tipgos 1981).

Another research approach considered more applicable to this study centers on the concept of developing knowledge clusters based on higher levels of abstraction as the field of international business develops and matures. A significant literature in the area of international management attributes has been developed on the basis of clustering methodology and some applications to international accounting and auditing have been recognized. Choi and Mueller (1978) identify a specific six-step process for application to international accounting:

1. description of phenomena observed;
2. systematic classification of observations according to a generally useful taxonomy;

3. comparisons and analyses of the observed phenomena according to classes and subclasses;
4. abstraction of general characteristics and principles from the analyses undertaken;
5. determination of relatively few basic concepts underlying the evolving field of inquiry; and
6. theory (model) building and testing, congruent with the foregoing developmental step and aimed at predictive processes.

This type of research in the area of comparative management has attempted to establish clusters of countries based on similarities of relevant organizational variables. A project by Ronen and Shenkar (1985) reviews eight empirical studies that use employee attitudinal data (primarily low and middle management level employees) to group countries and synthesize a cross-cultural clustering map based on integration of the eight studies. The eight empirical studies reviewed by Ronen and Shenkar were: Haire, Ghiselli, and Parker (1966), Sirota and Greenwood (1971), Ronen and Kraut (1977), Hofstede (1976), Griffeth, Hom, Denisi, and Kirchner (1980), Hofstede (1980), Redding (1976), and Badawy (1979).

This paper combines the clustering concepts from management research with attitudinal data collected via nominal group techniques conducted with internal auditors of U.S.-based MNCs in order to identify key factors that explain variance in auditor attitude and audit quality in all audit engagements plus the added problems and dimensions of international audits.

PROBLEMS IN INTERNATIONAL AUDITING

Two of the three fundamental obstacles to international auditing, as identified by Needles (1989), include the lack of a theory or framework of international auditing and the failure to understand how cultural differences affect the audit process. These same two general categories can also be used to reflect the practical problems facing internal and external auditors who are born and trained in the United States and temporarily assigned to the audit of MNCs or subsidiary units in another country. Although similar problems may exist for the auditors from other countries performing audits within the United States, this paper is limited to identification, explanation, and weighting of the variables affecting U.S. auditors' performance of international audits.

Culture has been defined as "a learned, shared, compelling, interrelated set of symbols whose meanings provide a set of orientations for members of a society. These orientations, taken together, provide solutions to problems that all societies must solve if they are to remain viable" (Terpstra and David 1985). Characteristics that tend to be constant within a given culture, but vary across different cultures, can be used to cluster countries according to culture. One text has identified these characteristics as: (1) communication and language, (2) dress and appearance, (3) food and feeding habits, (4) time and time consciousness, (5) rewards and recognitions, (6) relationships, (7) values and norms, (8) sense of self and space, (9) mental process and learning, and (10) beliefs and attitudes (Harris and Moran 1979). Terpstra and David (1985) reduce this list to four subsets of culture which generate potential problems in the conduct of business—status, custom, language, and law. Each of these four will be discussed in terms of potential auditor problems.

An individual's or group's status affects the relationship of that individual or group to other individuals or groups. Status also confers certain rights and obligations on the incumbents. In some cultures social and occupational status is mobile (subject to change), but is constrained in other cultures. Auditors must be aware of an individual's or group's status in order to determine the level of understanding and degree of creditability associated with evidence obtained from that individual or group.

Custom provides an indicator of what may and may not be done within a society and is often associated with religion. A three-stage model of human action is presented in Figure 1 to illustrate variations in possible action predicated on the customs and religious beliefs of a culture. The same action constrained by one culture may be approved by another. For example, lying to or misleading a U.S. auditor is approved action in some Eastern cultures while eating with the right hand would be unacceptable.

Language may be used by an auditor to the benefit or detriment of the audit process. Prior to and exclusive of reporting, cross-cultural communications pose the potential of misunderstanding for both sides of the audit process. An obvious hurdle is sending and receiving accurate messages in a second language. Even within communications between relatively homogeneous members of the same culture, words may be emotive and carry parallel messages. Responses to what an international auditor intends to be a neutral statement or question may include strongly positive or negative emotions which inappropriately influence the message communicated.

Laws of different cultures have varied philosophical bases. Western law is based primarily on Greek philosophy which assumes members to be literate and have regard for other individuals. Islamic law is based on religious intuition rather than logical thought, and contains a permanent state of belligerence and pugnacity against nonbelievers. Black African law is based on nonliterate tribal oriented thinking while the Chinese concept of law is predominantly Confucian. Indianized Asian law contains metaphysical constructs based on scripture. A cross-cultural legal-based problem commonly stated by U.S. auditors is that compliance with the Foreign Corrupt Practices Act would prohibit U.S.-based MNCs from engaging in certain business practices considered legal, moral, and even necessary in many other countries.

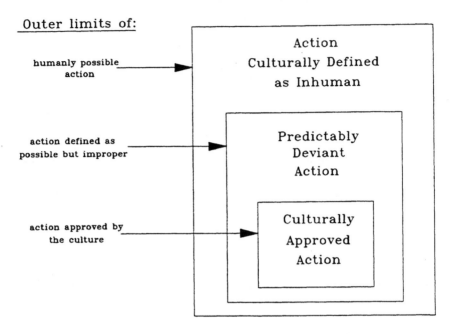

Source: Adapted from Terpstra and David (1985, 50).

Figure 1.   Limits to Human Action

APPLICATION OF THE CLUSTERING APPROACH

In one example of a clustering methodology applied directly to international auditing, it was found that audit reports can be divided into five groupings of countries with similar practices (Hussein, Bavishi and Gangolly 1986). In the current study, greater emphasis is placed on the audit process leading to the audit report and the attitudinal variables of both the auditor and auditee that impact the quality of the output. The previously referenced Ronen and Shenkar (1985) clustering map is presented in Figure 2 and serves as the basis for identification and discussion of variables potentially impacting the audit process as well.

The map in Figure 2 contains eight different clusters plus a group of four independent countries which represent most of the noncommunist world. The first, and most important, conclusion is that most countries/nations can be clustered according to culture based on four attitudinal variables—work goals, values, needs, and job attitudes. The discriminant validity of these four

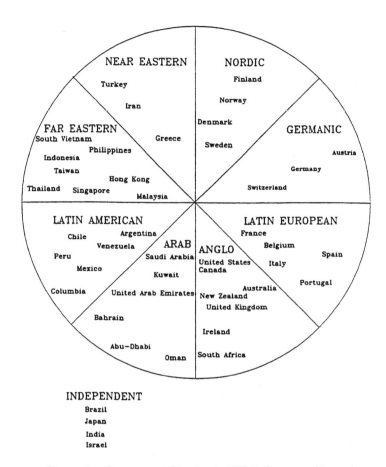

Figure 2. Ronen and Shenkar's (1985) Country Clustering

variables is supported by the synthesis of the eight empirical studies. The resulting clusters consistently discriminate between the cultures represented by different countries. The support for the Anglo, Germanic, Nordic, Latin European, and Latin American clusters is presented as very strong while the Far Eastern, Arab, and Near Eastern clusters along with the independents require additional empirical research to further identify substantial differences in the variables.

In addition to the grouping of countries within each cluster, Figure 2 illustrates per capita GNP as concentric distances from the center of the map. The most highly developed countries are closer to the center of the map and appropriately indicate more similar levels of development and attitudes even if not within the same cluster. For example, even though Switzerland and

Sweden are in separate clusters, similarities in economic development and technology create a closeness in attitudes that override many other differences. Alternatively, Columbia and Finland are in separate clusters and are mapped as being significantly different without advanced economic development considerations creating an overriding closeness. The limitations of a two-dimensional scaling do not allow for indications of intercluster similarities other than per capita GNP. This limitation is amplified for the four countries classified as independent because the analysis of attitudinal variables did not permit grouping with any of the other clusters, nor with one another. One hypothesis for the four countries (Israel, Japan, Brazil, and India) that are classified as independent of the other cluster is that in these countries, professionals place sufficient importance on economic and technological development to cause separation from the more common geographic grouping. If a three dimensional map were available, the independent countries would be clustered near the center of the map on the third dimension.

A better understanding of why certain countries cluster, as exhibited in the map, can be obtained from looking at other dimensions underlying the clusters. Ronen and Shenkar identify and discuss three such interdependent dimensions in addition to the economic and technological development—geography, language, and religion. It can be generally observed that the countries with one of these elements in common often share all three and they can serve to clearly discriminate between clusters. For example, the Latin American cluster is geographically contiguous with most residents of the member countries sharing the same language and religion. It can also be observed that people in Anglo countries speak English and people in Germanic countries speak German. With respect to religion, the Anglo, Germanic, and Nordic cluster are predominantly Protestant while the Latin clusters are mostly Catholic. Far Eastern countries share Eastern religions.

Although countries in each cluster mostly share commonalities in geography, language, and religion, some exceptions can be explained. The Anglo cluster shares a common language and is fairly homogeneous in religion, but has tremendous geographic separation. In this case the cluster contains countries from all five continents but all can be identified as cultural extensions of British colonization. The Far Eastern cluster contains countries geographically close but separated by oceans and quite heterogeneous with respect to language and religion. An observable cultural similarity, however, is that each of the countries has approximately equal mixtures of similar languages or religions.

Application to International Audit Quality

Findings of the eight individual empirical studies synthesized into the previously discussed cross-cultural mapping also provide significant practical advantage when attitudinal similarities can be identified by country. Haire et al. (1966) conclude that approximately one-third of the variance in work goals and managerial attitudes could be explained purely by country of residence. Similarly, England (1978) and Griffeth et al. (1980) find that approximately one-half of work goal and job attitude differences are explained by country differentiation. With this information, managers in MNCs can better understand employee attitudes, establish compatible regional units, effectively place international management assignees based on goals, and generally improve results of policies and practices based on national boundaries.

This study is based on the premise that cross-cultural attitudinal differences of auditors and auditees can be identified and clustered by country in a manner similar to the Ronen and Shenkar grouping of managerial attitudes. If cross-cultural variables or factors that affect the effectiveness and efficiency of international audits can be identified, then this increased understanding would permit improvement in audit engagement quality via assignment of the engagement team personnel, selection of audit procedures, and between engagement comparisons of the identified cross-cultural variables.

The methodological difference is that, unlike the management mapping project, multiple empirical studies which have collected attitudinal data on similar variables across many different countries are not known to exist for international audits. Accordingly, this study seeks to collect such data from U.S.-based internal auditors who have experience in performing audits in multiple countries for their MNCs. To the extent that experienced auditors' perceptions of the factors impacting audit quality accurately reflect a comprehensive set of cross-cultural attitudinal variables existing in other countries, a proper clustering of countries based on the multidimensional variables can be generated. The remainder of this paper describes the methodology of obtaining the auditor determined quality factors, constructing a dimensional model utilizing the identified quality factors, and comparing the resulting model with the cluster mapping synthesized from the empirical studies of managers' attitudinal variables.

RESEARCH METHOD

In order to move toward an understanding of how cross-cultural factors affect the audit process during international engagements, it was decided that the auditors themselves would likely provide the most insight. The research method applied in this study is based on a specialized form of the nominal group techniques developed by Adam, Hershauer, and Ruch (1978, 1986) specifically for use in understanding the factors affecting performance quality for service industries. Nominal group techniques are premised on the belief that the people involved in the day-to-day completion of the process under study can provide insight into weaknesses and potential problem areas within that process. The specific techniques prescribed by Adam et al. have been successfully applied in several banks (Adam et al. 1986; Aubrey and Eldridge 1981), the federal reserve system (Adam et al. 1986), corporate management groups (Theby 1981), systems development groups (Hershauer, St. Louis, and Green 1984), and external auditing (Sutton and Lampe 1991). In this study the modified nominal group techniques were applied with two groups of experienced internal auditors from two different U.S.-based oil companies that rely on their U.S. trained internal auditors to perform fieldwork for virtually all the audits of their international units.

Nominal Group Techniques

The nominal group techniques used in this study are an adaptation of the procedures developed by Adam et al. (1986) and used by Sutton and Lampe (1991) in their study of the factors affecting quality in domestic external audits. Full application of the techniques would achieve four objectives: (1) definition of the audit process, (2) identification of the critical factors affecting the audit team's ability to achieve the level of audit quality desired, (3) generation of measures for each of the quality factors, and (4) evaluation of the relative affect of each of the critical factors on overall audit quality. For purposes of this study, the focus was necessarily on the first, second, and fourth objectives. Additionally, the first two objectives were examined from two perspectives—first for domestic factors and then for additional international factors (see Table 1).

During the first session, the objectives of the process and the nominal group techniques to be applied were first discussed with the participating audit teams. The remainder of each session was spent on arriving at a consensus as to how the audit process would be defined for purposes of a U.S. audit engagement. The "defining" process focused on defining where the audit process was considered to start and end. The set of boundaries established the limitations for study during each of the succeeding sessions.

During the second session, the defined audit process was divided into a series of sequential stages. In past studies, improvements in task analysis have been found when the tasks are separated and studied in smaller parts (Adam et al. 1986). Additionally, the success of succeeding sessions is dependent on defining stages that represent separate and distinct subprocesses within the overall audit process. A general rule in defining a stage is that there should be a definable input into each stage, and that the output of a prior stage triggers the beginning of the following stage. The conclusion of this session was again marked by group consensus on a set of sequential stages representing the defined audit process.

The third session focused on identifying the factors having a critical effect on audit process quality. An audit quality factor was defined to the participants

Table 1. Summary of the Nominal Group Technique

Session	Process	Technique	Description
1	Defining the boundaries of the audit process	Group Consensus	*Defining the boundaries* focuses on identifying the scope of the audit process where the key factors affecting the auditors ability to perform the desired level of quality are likely to exist.
2	Identifying the sequential stages in the audit process	Group Consensus	*Sequential stages* are segments of the defined audit process in which a definable input enters the segment, a state change occurs during throughput, and a definable output exits the process.
3	Idenifying key domestic audit quality factors	Structured Brainstorming Q-methodology	*Audit quality factors* represent the variables in the audit process affecting the relationship between auditor effort and attainment of audit objectives.
4	Re-defining the audit process by international dimensions	Group concensus Q-methodology	*Re-defining* the process focuses on how different country affects can be clustered over a primary set of dimensions representing the impact on the audit process.
5	Identifying key internatinal audit quality factors	Structured Brainstorming	*International audit quality factors* represent additional variables that affect total attainment of audit objectives and are associated with cross-cultural dimensions.
6	Evalualting the relative impact of audit quality factors	Individual Ratings	*Magnitude measurement scaling* provides a method for identifying the relative impact of each factor.

as any variable affecting the audit team's ability to achieve the level of audit quality desired. The identification of audit quality factors, and in turn the key factors affecting audit quality, consisted of a four step approach.

1. A structured brainstorming technique was initiated where each participant independently generated a list of all factors that may affect an auditor's performance.
2. Through a round-robin sequence, an aggregate list of factors was formed by taking one factor from each participant's list during each round.
3. After exhausting all of the participant's lists, the audit team reflected on the combined list, and steps 1 and 2 were repeated to list any additional factors that were identified.
4. The combined list of audit quality factors was evaluated independently by each of the participants using Q-sorts to divide out the perceived critical factors.

The analysis of the data in step 4 is based on Q-methodology techniques (Kerlinger 1973). The first step in the analysis is the collection of Q-sort data. A Q-sort consists of separating items into categorical groups. In this case, the participants first divided the list of factors into factors they felt had a significant affect on overall audit quality and those that they considered relatively unimportant. Each participant then takes his/her list of significant factors and ranks them in order of importance. The final step consists of aggregating the participants' Q-sorts into a composite list of key audit quality factors, and a list of relatively unimportant factors that are dropped from the remainder of the study. The quality factor identification process is repeated for each audit process stage.

During the fourth session, the focus shifted to examining how performing an internal audit in an international location influences the quality of the audit process. The same structured brainstorming approach was used to identify major shifts in the audit process that take place due to changes in environment at different international locations. Discussion was then directed to how different cross-cultural impacts can be used to cluster countries according to a primary set of dimensions representing an impact on the entire audit process—essentially "re-defining" the audit.

The fifth session was a replication of the third session, except focusing on what additional factors have critical affects on audit process quality when performed in an international environment. The identification of "international" audit quality factors used the same four-step approach as used in session three: (1) structured brainstorming derivation of all factors affecting audit quality, (2) round-robin listing of individually generated factors into a combined list, (3) reflection on the total list to identify any ignored factors, and (4) application of Q-sorts to separate the key audit quality factors from the relatively unimportant factors.

The sixth and final session evaluates the relative impact of each factor on the audit process. This was accomplished by using a magnitude measurement scale designed to elicit the relative impact. To facilitate the magnitude measurement process, one factor is selected for its commonality and ease of understandability and is assigned a value of 100. This factor is then used as a benchmark from which to evaluate the remaining factors. For this study, the identified audit quality factor "Adequacy of Working Papers" was selected as the benchmark for relative evaluation of all other quality factors. If another factor was perceived to be three times as important as the benchmark factor, a value of 300 (3×100) would be assigned. On the other hand, if a third factor being rated was perceived to be only one-half as important to overall audit quality as the benchmark, then the factor would be assigned a value of 50 (.5 $\times$ 100). The geometric mean—calculated by taking the log of the magnitude measures, averaging the logs, and taking the antilog of the mean of the logs—

is subsequently used to aggregate the responses into a single weighting factor (Howard 1981; Howard and Nikolai 1983; Sutton and Lampe 1991).

Participants

In selecting participants for the nominal group techniques, Adam et al. (1986) note that an optimal group size and mix includes 6 to 12 participants representing all levels of employees involved directly with the process being measured. In this study, the first group was formed from the internal audit staff of a large natural gas company. The group represented an audit team consisting of two managers and four seniors. While the audit staff levels for this firm consist of manager, senior, and staff, the firm has not had anybody at the staff level for several years. Experience ranged from 5 to 17 years with the company and 1½ to 6 years of internal audit staff work within the company.

Replication of the nominal group technique with a second group of auditors was considered desirable as a means of evaluating the validity of the cross-cultural dimensions identified during the "re-defining" stage of the study. A second group of internal auditors from a different international oil and gas company provided the desired validation. This audit team also included 6 auditors (1 associate, 4 seniors, and 1 supervisor). Most auditors enter this company with prior external audit experience and leave the internal audit department within 3 to 5 years to take other positions in the organization. Therefore, virtually all performing staff auditors are at one experience level. Experience ranged from 1 to 10 years with the company and 4 to 15 years total audit experience.

All of the participating auditors from both companies had significant prior international audit engagement experience including engagements in Australia, Ecuador, France, Great Britain, Indonesia, Japan, and Venezuela. The variety of experience included countries from most of the previously identified clusters derived from managers' attitudinal data.

RESULTS

Presentation of the results generated via the nominal group techniques is divided into two sections. The first section will present an overview of "defining" the audit process within domestic environments (e.g., United States), and then "re-defining" this environment when the audit fieldwork is performed outside the United States. The second section presents the audit quality factors generated and examines differences in factor affects on domestic versus international audit engagements.

"Defining" the Audit Process

The "defining" of the audit process consisted of sessions one, two, and four from Table 1. The first session focused on setting the boundaries for the beginning and end of the audit process; the second session was oriented to breaking the audit process into stages; and the fourth session concentrated on the impact dimensions of moving the audit process to an international environment.

Through a consensus decision approach, the internal auditors defined the audit process as beginning with the selection of an entity for audit in a given year. The end point of the audit was determined to be the issuance of the final audit report to the auditee. The setting of these boundaries through group consensus was designed to assure that all stages of the audit process having a significant impact on overall audit engagement quality were included in the "defined" audit process.

During the second session, the "defined" audit process was divided into a series of sequential stages. Again using a group consensus approach, the audit process was divided into three distinct, sequential stages: (1) Preliminary Survey, (2) Fieldwork, and (3) Reporting. These three stages were considered the lowest level of segmentation of the audit process that could be devised while maintaining a series system (i.e., completion of one stage leads to the beginning of the next). The boundaries and sequential stages identifying the audit process are not terribly surprising—being readily paralleled with similar findings in the external audit environment where the primary stages have been identified as (1) Planning, (2) Fieldwork, and (3) Final Administration (Sutton and Lampe 1991). However, this phase of the nominal group technique was important to setting up a framework from which to complete the remaining parts of the study.

The fourth stage was subsequently used to provide insight into the major cross-cultural dimensions affecting audit quality as audits take place in different countries. An open format discussion was used to allow the participating auditors to discuss how they each perceived the audit environment to change as they worked in other countries. After discussion started approaching a consensus view of how the environment changed, the group was asked to deduce the major dimensions impacting these changes when audits were conducted in different countries. The group concurred on three primary dimensions that included most of the cross-cultural variables impacting audit quality in different countries: (1) non-English versus English speaking, (2) Eastern versus Western culture, and (3) third world developing versus economically developed countries. These three dimensions are presented in Figure 3 in terms of an "impact cube for cross-cultural affects." Replication of this phase with the second group of internal auditors yielded verification of the three dimensions with the six auditors from the oil and gas company deducing an identical consolidation of dimensions.

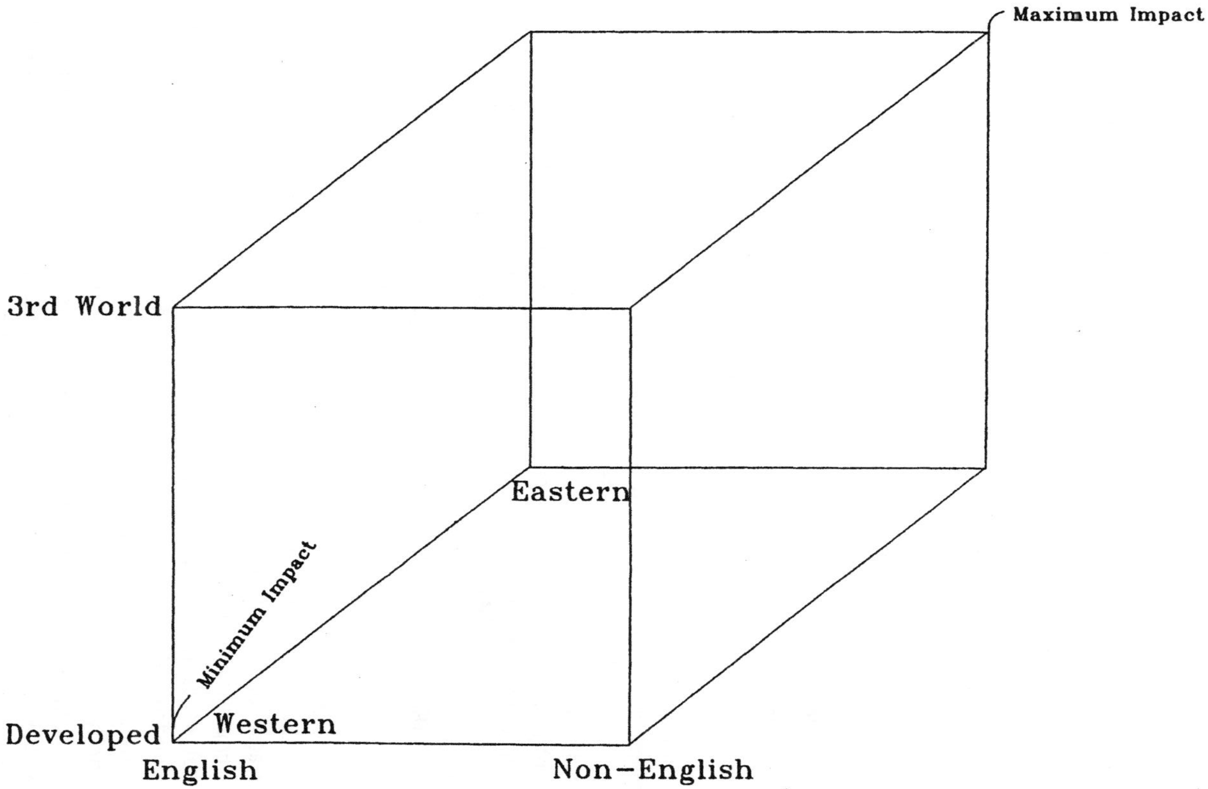

Figure 3. Impact Cube for Cross-Cultural Affects

An important feature of the impact cube generated by auditors and presented in Figure 3 is the similarity of dimensional impacts with those in the Ronen and Shenkar (1985) synthesis of eight empirical studies of international managers. It may be recalled that the two-dimensional mapping included a dimension of economic and technological development that crossed the cluster boundaries on the map and was hypothesized as a primary reason for the four independent countries. Additional dimensions of language, geography, and religion were noted as relevant to and consistent with the identified clusters although not illustrated on the two-dimensional map. These empirically derived cross-cultural dimensions can be compared with the three dimensional impact cube in Figure 3 as follows:

Ronen and Shenkar	*Impact Cube*
Economic and technological development	Developed vs. Third World Developing Countries
Language	English vs. Non-English
Geography/Religion	Western vs. Eastern Philosophy

The combination of geography and religion is not considered a significant difference because in Ronen and Shenkar the only exception to a linkage of geography with language and religion was the Anglo cluster in which British colonization appeared to be the overriding cultural impact transcending geography. Because this study is based on U.S. auditors performing fieldwork in other countries, the Anglo exception to the geography dimension is eliminated and linkage between geography and religion is not considered a limitation.

Due to the consistency between the cross-cultural dimensions generated by U.S. auditors and those empirically derived from multinational managers, additional validity is accorded utilization of the impact cube to help identify deviations in international audit quality. From the most basic perspective of analysis, the further away an auditor is on any combination of cube dimensions, the greater the chance of reduced audit quality due to cross-cultural impacts on the engagement. It must be noted, however, that each dimension is ordinal and not necessarily equally important to audit quality. This means that deviation from the origin is not quantitatively additive on a ratio scale—that is, two units of deviation from English plus one unit of deviation from Western philosophy does not equal three total deviations from the origin. The ordinal limitation, however, does not diminish many of the benefits of using the impact cube to consider effects on audit quality. The impact cube provides a consistent framework on which to base the conclusion that an international audit with fieldwork performed in Australia carries much less cross-cultural impact on quality than an audit engagement in Thailand because the amount of deviation on each dimension (English language, developed country, and Western philosophy) is clearly less in Australia than in Thailand.

Additional precision in determining the amount of cross-cultural impact is attainable via the weighting of each dimension. The auditors participating in this study clearly considered the English vs. non-English dimension as the most important to total audit engagement quality. The greater the degree of difficulty in English communications with auditees, the lesser the ability to efficiently obtain audit evidence with the degree of confidence needed to attain the audit competency objective. In addition to the pure language barrier, auditors perceived that auditees used nonEnglish language excuses to intentionally withhold information even to the extent of misleading the auditor. This attitude was interrelated with the dimension considered second in importance— deviation from Western philosophy.

The degree of deviation by an auditee from adherence to Western philosophy was perceived to be directly correlated with problems in maintaining a necessary level of understanding and relationship between auditor and auditee. For example, several common behaviors in U.S. culture are considered offensive and/or insulting in Eastern culture (e.g., showing the soles of your shoes while sitting in a meeting, pointing at a person, etc.). Further, simply selecting the audit team becomes much more important as a gray-haired male will immediately command respect while a female will tend to receive little, if any, respect. In general, failure to understand and comply with such cross-cultural differences is likely to result in a less effective and less efficient audit. The greater the deviation from Western philosophy, the greater the difficulty for U.S. auditors to complete a quality audit.

The third world dimension of the impact cube was considered important to audit quality but less so than deviations from the English language and Western philosophy. One reason for this ranking of importance is that the auditors perceived problems generated by performing fieldwork in lesser developed countries to be more personal than professional. Microcomputer technology coupled with data base, word processing, and other such support systems usually permit the application of advanced technology audit procedures in less developed countries as well as in the United States. Although the personnel, accounting procedures, and record maintenance in third world countries present some problems, the effect on audit team morale when living conditions deteriorated was considered more critical. Concerns such as health maintenance, emergency procedures, food preparation, and communication links with home become more important to the auditors.

The advantage of a consistent framework, such as that provided by the impact cube in Figure 3, is that the degree to which cross-cultural variables are likely to affect audit quality can be predicted, planned for, and mitigated to some extent. In order to be more specific as to the extent to which each dimension has on a specific engagement, it is necessary to identify and weight the importance of factors causing deviation from the origin for each dimension. The following sections explain how such audit quality factors were identified and weighted.

Audit Quality Factors

As previously noted, the identification of audit quality factors consisted of a four-step approach: (1) structured brainstorming derivation of all factors affecting audit quality, (2) round-robin listing of individually generated factors into a combined list, (3) reflection on the total list to identify any ignored factors, and (4) application of Q-sorts to separate the key audit quality factors from the relatively unimportant factors. This process was used in session three to generate domestic audit quality factors and again in session five to generate additional international factors.

During session three, the generation process led to a list of 75 distinct audit quality factors for the three stages of the audit. The Q-sorting application led to a refinement of the list to a set of 17 key audit quality factors (see Table 2).

Table 2.   Selected Audit Quality Factors

Domestic Audit Quality Factors	International Audit Quality Factors
Preliminary Survey	
Management Communication of Scope and Objectives	Travel Expenses and Limited Resources
Availability of Auditee Data and Personnel	Language Barrier
Surveyor's Professional Proficiency	
Auditor's Time Constraints	
Clarity of Scope and Objectives	
Fieldwork	
Adequacy of Working Papers	Language Barrier
Level of Performance of Audit Procedures	Cultural Practices
Audit Team Communications	Customs and Holidays
Auditor's Time Constraints	Different Accounting Methods and Systems
Quality of Audit Programs	Health Risks
Auditor's Professional Proficiency	Burnout
Level of Auditee Cooperation	Audit Team's International Experience Level
Audit Team's Cohesiveness	International Political Environment
	Equipment
	Expenses
Reporting	
Sensitivity of Findings (political)	Equipment Functionality and Compatibility
Auditor's Professional Proficiency	Distance Factors for Review Meeting and Responses
Auditor Time Constraints	
Level of Auditee Cooperation	

Table 3.   Geometric Means for Audit Quality Factors

Audit Stage	Cross-Cultural Location Type			
	Domestic	Eastern Culture	Third World	Non-English
Preliminary Survey				
D Management Communication of Scope and Objectives	86	89	89	89
D Availability of Auditee Data and Personnel	99	122	122	122
D Auditor's Professional Proficiency	74	88	81	84
D Auditor Time Constraints	68	62	62	62
D Clarity of Scope and Objectives	86	112	115	115
I Travel Expenses and Limited Resources	14	77	77	77
I Language Barrier	1	110	114	114
Fieldwork				
D Adequacy of Working Papers	100	100	100	100
D Level of Performance of Audit Procedures	150	144	144	144
D Audit Team Communications	131	155	155	155
D Auditor Time Constraints	129	128	128	131
D Quality of Audit Programs	104	114	114	114
D Auditor's Professional Proficiency	144	148	148	148
D Level of Auditee Cooperation	100	153	153	153
D Audit Team Cohesiveness	96	141	141	141
I Language Barrier	1	197	206	206
I Cultural Practices	2	87	93	87
I Customs and Holidays	1	66	66	66
I Different Accounting Methods and Systems	13	83	189	88
I Health Risks	2	53	173	53
I Burn Out	9	58	67	67
I Audit Team International Experience	1	108	121	121
I International Political Environment	1	58	67	67
I Equipment	8	57	61	61
I Expenses	10	79	84	79
Reporting				
D Sensitivity of Findings (political)	52	87	90	87
D Auditor's Professional Proficiency	103	113	108	108
D Auditor Time Constraints	67	78	78	82
D Level of Auditee Cooperation	67	128	136	136
I Equipment Functionality and Compatability	6	54	54	54
I Distance Factor for Review Meeting and Response	6	111	117	117

Note: D = domestic figures; I = international factor weightings.

Completion of the generation and evaluation process for international audit quality factors led to an additional 14 key audit quality factors being added to the list for international environments (see Table 2).

The total of 31 quality factors listed in Table 2 represent the key factors identified to impact audit quality in both domestic and international audits. The listing, however, does not indicate the relative degree of importance or impact each factor has on total audit quality. Although it may change slightly from one audit to another, it is important to estimate the relative affect of each domestic and international factor.

The relative impact of each factor was elicited in session six using the magnitude measurement approach. "Adequacy of working papers" was selected as the benchmark factor and assigned a value of 100. The participants were then asked to provide a magnitude measure for each of the other factors relative to the benchmark. Separate weighting factors were assigned for a factor's impact in (1) domestic audits, (2) audits in Eastern culture countries, (3) audits in third world countries, and (4) audits in non-English speaking countries. The magnitude measures are listed in Table 3 with the domestic figures listed first (preceded by a "D") and the international factor weightings preceded by an "I."

Interpretation of the geometric means is based on the common benchmark—"Adequacy of working papers." Under the domestic column it may be noted that all of the 17 domestic quality factors range from a low of 52 (about one-half as important as the benchmark) for sensitivity of the reporting to a high of 150 for the level of audit procedure performance. For domestic audits, however, the 14 cross-cultural international factors have very little importance. Under the three columns representing the key dimensions of international audits, all of the 31 quality factors range in importance from 53 to 206.

The most obvious comparison of relative quality factor weightings involves the language barrier. Although virtually nonexistent in domestic audits, it is clearly the most important fieldwork factor in all three dimensions of international audits. It should also be noted that most of the factors identified as important to domestic quality are also perceived to have significant impact on international audits. Some factors important in domestic audits became even more important in international audits. For example, "availability of auditee data and personnel" during the preliminary survey phase of an audit and the "level of auditee cooperation" in both the fieldwork and reporting phases become more important to the end quality of international audits.

In further examining Table 3, it can be observed that the primary additional factors having impact in international engagements are:

1. the language barrier (preliminary survey and fieldwork),
2. clarity of scope and objectives (preliminary survey),
3. audit team cohesiveness (fieldwork),
4. audit team international experience (fieldwork), and
5. distance factor for review meeting and responses (reporting).

Additionally, for the third world countries, two other factors significantly increased in import:

1. different accounting methods and systems (fieldwork) and
2. health risks (fieldwork).

SUMMARY AND CONCLUSIONS

The audit profession has been faced with two major challenges in recent years that will likely continue to increase in the foreseeable future—(1) an increasing demand for quality service performance, and (2) the complexity of auditing in a multinational corporation. Beyond the traditional internal and external pressures on the audit profession for better quality, the multinational corporate environment is making scrutiny by one country's auditors of another country's businesses a common and necessary task. To understand how to improve audit quality in this complex environment, better understanding of the audit process and the cross-cultural influences is needed.

Similar problems in understanding the environment and how to interact with business employees in international locations have been studied by management researchers. The management research has focused on using clustering methodologies to group countries by common characteristics and devise a framework from which to study cross-cultural influences in the business environment. The comparative similarities with the objectives of this study have led to a similar approach in studying the affects of different international settings on audit quality.

In this study, nominal group techniques were used to structure small group interaction with internal auditors from a large international gas corporation. The structured interaction focused on discussion of how different international locations affected their ability to perform the level of quality audit desired. The auditors subsequently worked on categorizing the differences into major dimensions by which to view the audit quality impact of various international settings. Three dimensions evolved as the dominant cross-cultural influences: (1) English versus nonEnglish speaking, (2) Western versus Eastern culture, and (3) developed versus third world countries. These dimensions match very closely with the dimensions found in prior management studies (see Ronen and Shenkar 1985).

This study also moved beyond simply identifying the major cross-cultural forces to examining how these cross-cultural forces change the factors having primary affect on the auditors ability to achieve a desired quality level. The changes in factors due to cross-cultural forces were examined via a magnitude measurement scale that captures the relative affects of different factors. Study across each of the cross-cultural dimensions revealed significant changes in many factors, with several factors considered essentially unimportant in the domestic arena becoming some of the most important factors in different international settings.

The findings presented in this study are important for quality improvement in the audit process. While the most critical stage in terms of international factor impact is fieldwork, the main factors rising to the height of importance could be mitigated to some degree during the selection and preparation of the audit team during the planning stages (part of the Preliminary Survey). This will require the audit planner to carefully consider each of the dimensions that impact a particular engagement and then use international experience and judgment with respect to the factors within each dimension to determine how to best mitigate the negative impact of each identified dimension and how to best maximize the positive influences on audit quality.

The framework presented in this study provides a basic level of understanding in how the international environment affects the audit process. As noted by Needles (1989), additional research is needed to expand the full understanding of the international audit environment. While this study focuses on the cross-cultural dimensions of the environment and the related impact on factors affecting the ability of auditors to achieve a desired quality level, objective measures of the change between engagements in the presented audit quality factors would further enhance the audit planners ability to counter the affects. Additionally, while the impact cube integrates the three identified cross-cultural dimensions, further understanding of the compound affect from changes in multiple dimensions of the cube is needed. An understanding of the compound affects of dimension changes would be a significant aid to the audit planner.

REFERENCES

Adam, E. E., Jr., J. Hershauer, and W. Ruch. 1978. *Measuring the Quality Dimension of Service Productivity*. National Science Foundation, Grant No. APR 76-07140. Washington, DC: U. S. Department of Commerce, NTIS, PB282243/AS.

————. 1986. *Productivity and Quality Measurement as a Basis for Improvement*. University of Missouri-Columbia Business Research Center.

Aranya, M., and J.H. Armenic. 1981. Public accountant's independence: Some evidence in a Canadian context. *Journal of Accounting Education* (Spring): 11-17.

Aubrey, C.A., and L.A. Eldridge. 1981. Banking on high quality. *Quality Progress* (December): 14-19.

Badawy, M.R. 1979. Managerial Attitudes and Need Orientation of Mid-Eastern executives: An empirical cross-cultural analysis. Paper presented at the meeting of the Academy of Management, Atlanta. (August)

Bindon, K.R., and H. Gernon. 1987. International accounting research. In B.N. Schwartz et al. (eds.), *Advances in Accounting*, Vol. 4. Greenwich, CT: JAI Press, 43-65.

Buckley, J.W., and P.R. O'Sullivan. 1980. International economics and multinational accounting firms. In J.C. Burton (ed.), *The International World of Accounting*. London: Arthur Young Professors' Roundtable, 115-37.

Choi, R.D.S., ed. 1981. *Multinational Accounting: A Research Framework for the Eighties*. Ann Arbor, MI: UMI Research Press.

Choi, F.D.S., and G.G. Mueller, 1978. *An Introduction to Multinational Accounting*. Englewood Cliffs, NJ: Prentice-Hall.

Dykxhoorn, H.J., and K.E. Sinning. 1981. The independence issue concerning German auditors: A synthesis. *The International Journal of Accounting Education and Research* (Spring): 163-81.

England, G.W. 1978. Managers and their value systems: A five country comparative study. *Columbia Journal of World Business* 13, no. 2: 35-44.

Griffeth, R.W., P.W. Hom, A. Denisi, and W. Kirchner. 1980. A multivariate, nultinational comparison of managerial attitudes. Paper presented at the meeting of the Academy of Management, Detroit (August).

Gul, F.A., and T.H. Yap. 1984. The effects of combined audit and management services on public perception of auditor independence in developing countries: The Malaysian case. *The International Journal of Accounting Education and Research* (Fall): 95-107.

Haire, M., E.E. Ghiselli, and L.W. Porter. 1966. *Managerial Thinking: An International Study*. New York: Wiley.

Harris, P.R., and R.T. Moran. 1979. *Managing Cultural Differences*. Houston: Gulf Publishing.

Hershauer, J.C., R.D. St. Louis, and G.I. Green. 1984. *Measurement of Systems Development Productivity, Technical Report 84-7*. Arizona State University Decision Systems Research Center.

Hofstede, G. 1976. Nationality and espoused values of managers. *Journal of Applied Psychology* 61: 148-55.

———. 1980. *Culture's Consequences: International Differences in Work Related Values*. Beverly Hills: Sage.

Howard, T.P. 1981. Attitude measurement: Some further considerations. *Accounting Review* (July): 613-21.

Howard, T.P., and L. Nikolai. 1983. Attitude measurement and perceptions of faculty publication outlets. *Accounting Review* (October): 765-76.

Hussein, M.E.A., V.B. Bavishi, and J.S. Gangolly. 1986. International similarities and differences in the auditor's report. *Auditing: A Journal of Practice & Theory* (Fall): 124-33.

International Federation of Accountants, International Auditing Practices Committee. 1983. *The Auditor's Report on Financial Statements*, Audit Guideline No. 13. New York: IFAC.

Kerlinger, F. 1973. *Foundations of Behavioral Research*. Toronto: Holt, Rinehart, and Winston.

Lee, M. 1978. The international auditor. *Internal Auditor* (December), 43-47.

Mann, R.W., and D.H. Redmayne. 1979. Internal auditing in an international environment. *Internal Auditor* (October): 49-54.

Most, K.S. 1988. *International Auditing*. Toronto: Canadian CGA Research Foundation.

Mueller, G.G. 1979. The state of the art of academic research inmultinational accounting. In W.J. Brennan (ed.), *Internationalization of the Accounting Profession*. Toronto: Canadian Institute of Chartered Accountants, 83-97.

Needles, B.E., Jr. 1981. Comparative international auditing standards: An overview. *Comparative International Auditing Standards*. Sarasota, FL: American Accounting Association.

———. 1989. International auditing research: Current assessment and future direction. *The International Journal of Accounting* 24, no. 1: 3-34.

Nobes, C.W. 1983. An empirical analysis of international accounting principles: A comment. *Journal of Accounting Research* (Spring): 268-70.

Redding, G. 1976. Some perceptions of psychological needs among manager in South-East Asia. Pp. 338-343 in *Basic Problems in Cross-Cultural Psychology*, edited by Y.H. Poortinga. Amsterdam.

Ricchiute, D.N. 1978. Foreign corrupt practices: A new responsibility for internal auditors. *Internal Auditor* (December): 56-64.

Ronen, S., and A.I. Kraut. 1977. Similarities among countries based on employee work values and attitudes. *Columbia Journal of World Business* 12, no. 2: 89-96.

Ronen, S., and O. Shenkar. 1985. Clustering countries on attitudinal dimensions: A review and synthesis. *Academy of Management Review* 10, no. 3: 435-54.

Scott, G.M. 1980. *Eighty-Eight International Accounting Problems In Rank Order of Importance: A Delphi Evaluation*. Sarasota, FL: American Accounting Association.

Securities and Exchange Commission. 1987. *Internationalization of the Securities Markets*. Washington, DC: SEC (October).

Sirota, D., and J.M. Greenwood. 1971. Understand your overseas work force. *Harvard Business Review* 4, no. 3: 53-60.

Stamp, E., and M. Moonitz. 1982a. International auditing standards—Part I. *CPA Journal* (June): 24-32.

———. 1982b. International auditing standards—Part II. *CPA Journal* (July): 48-53.

Sutton, S.G., and J.C. Lampe. 1991. A framework for evaluating process quality for audit engagements. *Accounting and Business Research* (Summer): 275-88.

Tantuico, F.S., Jr. 1980. Problems in adopting and implementing modern auditing techniques in developing countries. *International Journal of Government Auditing* (October): 10-17.

Terpstra, V., and K. David. 1985. *The Cultural Environment of International Business*, 2nd ed. Cincinnati: South-Western Publishing.

Theby, S. 1981. *Group Productivity Performance Measurement*. St Louis: McDonald Douglas Corporation CSW Productivity Council Measurement and Reporting Group.

Tipgos, M.A. 1981. Potential liabilities in international accounting practice. *Journal of Accountancy* (April): 171-76.

Vinton, G. 1991. International audit in perspective: A U.S./U.K. comparison. *Internal Auditing* (Spring): 3-9.

Wallace, W.A. 1987. An overview of research in international accounting and likely approaches to future inquiry. *Management International Review* (March): 3-7.

PART V

INTERNATIONAL ACCOUNTING EDUCATION

TOWARD INTERNATIONALIZATION OF UPPER-LEVEL FINANCIAL ACCOUNTING COURSES

Robert Bloom, Jayne Fuglister, and Jeffrey Kantor

ABSTRACT

This paper deals with differences in generally accepted accounting principles (GAAP) in five countries and how those differences could be used to improve the quality of accounting education in the intermediate and advanced courses. Through a critical examination of accounting principles in various countries, students can develop a "cultural awareness" and gain a better understanding of our own accounting principles.

INTRODUCTION

For the "common body of knowledge" courses, the American Assembly of Collegiate Schools of Business (AACSB 1986, 28-30) recommends "exposure

Advances in International Accounting,
Volume 5, pages 239-253.
Copyright © 1992 by JAI Press Inc.
All rights of reproduction in any form reserved.
ISBN: 1-55938-415-8

to the international dimension in one or more areas." However, in a study commissioned by the AACSB, Porter and McKibben (1988) find that business schools are largely complacent and tend to ignore long-term trends and developments. They identify several areas that sorely need curriculum improvement, one of which is "international dimensions." Business schools should give far more than "lip service" to this area, and, according to Porter and McKibben, multinational companies should lobby for a greater emphasis on international business in the curriculum. The American Accounting Association Committee on the Future Structure, Content, and Scope of Accounting Education (1986) has also observed (1986, 180-1):

> The minimum [accounting] program would develop in students the capacities for inquiry, abstract logical thinking, and critical analysis; literacy, which includes writing, reading, speaking, and listening; understanding numerical data; historical consciousness; ...the study of values; ... *international and multicultural experiences.* ...[180-181, emphasis added].

This paper considers generally accepted accounting principles (standards) in five countries, why those principles differ, and how these differences could be used to enhance the intermediate and advanced accounting courses. The focus is causality. That is, what are the particular cultural factors that determine the methods of accounting measurement selected in various countries? Information concerning *specific comparative accounting principles including explanations of the differences between countries* (e.g., due to cultural factors) is not currently available. Moreover, there are no published articles or other studies attempting to integrate differing accounting principles across countries. Cultural differences between countries may well be manifested in widely different principles.[1]

As for infusing international material, or for that matter any new material, into the upper-level financial accounting courses, there is likely to be strong opposition on the part of veteran faculty. Three key problems of internationalizing our accounting courses are: (1) the lack of time to present additional material in courses already packed with subject matter, (2) the lack of knowledge on the part of the faculty about international accounting, and (3) the lack of suitable materials to use in the classroom. A strategy for overcoming such inertia is to stipulate that no new topics will be added to the courses. The point is to use the existing topic mix as a framework for providing international applications. Nearly every topic listed on the syllabi for these courses lends itself to some form of internationalization, which can be covered in terms of specifying how and why practices and financial statements in other countries differ from those in the United States. Supplemental paperback minitexts, handouts, and anecodotes can be used to convey the material. The specific material furnished in this paper is intended to be supplemental.

Intermediate and advanced accounting textbooks currently do not provide adequate coverage of international aspects of accounting. Through a critical examination of accounting principles and an analysis of the derivation of those principles in other countries, students can develop a "cultural awareness" and gain a better understanding of our own accounting principles. An international approach ought to encourage students to read and examine foreign financial reports within the context of the economic, political, and social structures of the countries in question, and, in so doing, to broaden their accounting horizons. Moreover, the international material should enrich these courses by giving them a new motif for integration of topics and issues.

Most, when chairperson of the international section of the American Accounting Association, was critical of the lack of curriculum internationalization in accounting programs (1989, 1-3):

> We must admit that few accounting programs recognize the fact that accountants now function in a global financial market and must deal with international accounting and auditing problems as practitioners and business executives. Examine our accounting textbooks and you will find that the only part of international accounting that receives detailed treatment is foreign currency translation.... A doctoral student at any university knows that to select an international topic for a dissertation is like admitting having an illness unless the subject can be disguised as an exercise in capital markets, information economics, or agency theory research.

> Circulating course outlines, recommending international accounting course content, and even promoting the study of international accounting by writing and publishing books and articles on the subject appear to have had little impact on the curricula. Workshops on the internationalization of the accounting curriculum have not been held often enough or with enough participants to have had a noticeable effect.

Internationalization of the upper-level accounting courses is not a simple endeavor. The best advice is probably to go slowly at first, perhaps to do it initially in one section of the course on an experimental basis. As faculty develop greater expertise in the international material, they can expand the international component in their courses.

In this study, five countries, selected because of significant differences in their cultures and principle-setting processes, are analyzed. It is assumed that accounting principles are generally derived from the economic, legal, political, and social systems prevailing in each country. Then a comparative study of generally accepted accounting principles in each country, emphasizing the reasons for differences and similarities in specific principles or standards, is presented. The key questions underlying the study are: Can cultural and environmental factors explain differences in specific principles between countries? Is harmonization of principles likely among these and other countries? The following countries and principles are considered:

Countries

1. France
2. Japan
3. The Netherlands
4. Sweden
5. United States

Principles

1. Asset valuation methods
2. Asset revaluation methods
3. Consolidated financial statements
4. Inflation accounting
5. Inventories
6. Depreciation
7. Research and development
8. Interest imputation

The study was conducted by:

1. an examination of current literature from the American Institute of Certified Public Accountants Professional Accounting in Foreign Countries Series (1987 and 1988) and from major CPA firms on accounting principles in different countries;
2. an examination of recent issues of international periodicals;
3. interviews by the researchers with the controller of a major company operating in three countries under consideration—France, the Netherlands, and Sweden—in order to ascertain the particular accounting principles in these countries and how these principles are applied in practice (see Table 1).

Table 1. Interview Guide

---

1. Describe your country's accounting principle-setting authority. Is it public? private? full-time? part-time? Who are its members? How are they appointed? Are they completely independent?
2. Is a conceptual framework used in the princple-setting process? If so, what is the specific nature of this framework, and how is it used?
3. To what extent does your country "copy" in whole or part the accounting principles of other countries? Is your country a leader or a follower in this regard?
4. To what extent does your country "copy" in whole or part the accounting principles of other countries? Is your country a leader or a followerr in this regard?
5. Explain why your principle-setting authority adopted the principle you currently have as opposed to an alternative principle for each of the following:

> Asset Valuation
> Asset Revaluation
> Consolidated Financial Statements
> Inflation Accounting
> Inventories
> Depreciation
> Research and Development
> Interest Imputation

Explain how and why your principles differ from the corresponding U.S. principles. How rigid or flexible are each of your above principles? Why?

THE SIGNIFICANCE OF CULTURAL INFLUENCE ON ACCOUNTING PRINCIPLES

Kroeber and Kluckhohn (1952, 81) define culture in terms of:

> patterns, explicit and implicit, of and for behavior acquired and transmitted by symbols, constituting the distinctive achievements of human groups, including their embodiments in artifacts; the essential core of culture consists of traditional (i.e., historically derived and selected) ideas and especially their attached values; culture systems may on the one hand be considered as products of action, on the other as conditioning elements of future action.

Culture has been defined by Hofstede (1980, 25) as: "the collective programming of the mind which distinguishes the members of one human group from another." Culture is a very broad and complex concept, an umbrella term incorporating a vast array of characteristics. The concept of culture is too imprecise to measure.

The differences in GAAP among the countries can be explained by such cultural factors as:

- the economic system,
- the political system,
- the legal system,
- the social system,
- the sources of capital and extent of foreign trade,
- the influence of taxation on accounting,
- the emphasis of the government on accounting, and
- the status of the accounting profession in each country.

Zeff (1985, 1) suggests the idea that cultural factors can determine accounting measurement methods:

> It is difficult to disentangle cultural from legal, economic, institutional, and historical explanations—they are all intertwined. Furthermore, in a climate in which international harmonization is encouraged, it is not easy to distinguish national cultural factors as explanations of purely domestic trends and developments from those that might explain a country's proclivity to import standard-setting approaches and accounting practices from other countries. In general, I would argue that accounting influence follows economic and political influence ..., although cultural factors may also be at the root.

Further, Gray states (1988, 4):

> In interpreting the results of ... research relating to culture, the influence of any change factors will also need to be taken into account, bearing in mind the existence of external influences arising from colonization, war, and foreign investment including the activities of multinational companies, and large international accounting firms.

OVERVIEW OF U.S. VERSUS
NON-U.S. ACCOUNTING ENVIRONMENT

Number of Accounting Standards

There are various general differences between standards in the United States and other countries. U.S. GAAP (generally accepted accounting principles) are tightly written regulations trying to cover every conceivable facet of a particular accounting issue. In most other countries, there are far fewer accounting principles, generally allowing accountants and auditors more latitude to exercise professional judgment. The limited number and looser nature of non-U.S. principles, not to mention the slower pace in developing them, often reflects the sense of trust that prevails in other countries considered in this study with respect to companies, accountants, and institutions in general (Bloom and Collins 1988).

Conceptual Framework

In contrast to the United States, there is no written conceptual framework in the other countries under study, although some recognize the French national chart of accounts as such a framework. Non-U.S. principle-setting authorities evaluate principles primarily on experience and knowledge of accounting and the perceived economic consequences of the principles. In the United States, principles are evaluated increasingly by reference to the conceptual framework.

That U.S. principles are more numerous and more detailed than the corresponding principles in other countries does not imply their superiority. Accounting principles are very much a function of the economic, social, political, and legal environment of each country—that is, its culture. Principles that are suitable in one country are not necessarily suitable in another country. The United States, in view of a relative lack of trust, has tightly codified accounting principles even though, paradoxically, it has an unwritten common law. France, by contrast, has a codified law, but few accounting principles. France does, however, have a detailed *Plan Comptable* (Chart of Accounts), which is intended to foster uniformity and thereby assist in government economic and fiscal policies as well as business decision making. French principles are quasi-legalistic in nature. The importance of a government, in France as an example, as the main user of accounting, has a pronounced effect on the nature of the principles in various countries. Nevertheless, a number of French companies have been applying U.S. GAAP in their published financial reports with the aim of attracting investors in world markets.

Conservatism

In Sweden there is considerably greater emphasis on conservatism than in the United States, as reflected by the following practices. Tooling expenditures are expensed immediately in Sweden, but capitalized in the United States. The equity method of accounting for long-term investments is not used in Sweden, in contrast to the United States. In Sweden, gains and losses caused by fluctuations in exchange rates are reflected in the income statement only during the period incurred. In the United States, if the functional currency used by a subsidiary abroad is not the United States dollar, these foreign currency gains and losses appear in stockholders' equity. However, in the few cases where the functional currency is the United States dollar, then foreign currency gains and losses would be shown in the income statement. In France, the Netherlands, and Sweden, traditionally there has been an emphasis on conservatism, as also in the Fourth Directive of the European Community (EC). Conservatism is also stressed in Japan.

Legal Environment

France and most other countries have legal codes. On the other hand, common law prevails in the United States and England (Cairns 1990). In France, Japan, and Sweden the influence of government policies on accounting is pronounced. Tax regulations and accounting principles are largely the same, which is known as *tax conformity*.

In the Netherlands, emphasis is placed in setting accounting standards on conveying "a true and fair view," on economic substance rather than legal form; that is, the approach to formulating accounting principles in the Netherlands is not legalistic as it is in France. In France, the *Plan Comptable* calls for a uniform chart of accounts and financial statements. Since 1985, there has been a law in France calling for consolidated financial statements because of the need to raise funds abroad.

In France and Sweden, because they are tax conformity countries, deferred taxes are not recognized as they are in the United States. Taxable income is the same as book income in these countries. The only Swedish required accounting standards are found in its Companies Act of 1975 and Accounting Act of 1976.

As Cairns (1990), Secretary-General of the International Accounting Standards Committee, perceptively points out:

> Within any particular country, different factors may come into play for different accounting issues. For example, there is a direct link between tax rules and accounting rules for the financial statements of separate legal entities in France. In consolidated financial statements, however, this link has been broken so that accounting treatments that are not permitted

by tax law, for example, the capitalization of finance leases, are permitted, if not encouraged, in the preparation of consolidated financial statements.

It is also often suggested that the influence of tax rules is a major factor in determining accounting rules in continental Europe, but not in the so called Anglo-Saxon countries. As I am sure you are well aware, tax rules do have a major impact on accounting in Anglo-Saxon countries—for example, with the use of LIFO in the United States. Similarly, it was often suggested that companies in the United Kingdom would be willing to present current cost financial statements if they were acceptable to the tax authorities.

It is also often suggested that the influence of laws generally is far greater in continental European countries than in the Anglo-Saxon countries. Again, there is a trend towards a greater influence of the form of transaction rather than their substance in some aspects of Anglo-Saxon accounting. For example, whereas the capitalization of finance leases involves an application of the concept of substance over form, numerous corporations seek to avoid such a rule by relying on the legal form of a particular leasing contract.

UNITED STATES VERSUS
NON-U.S. ACCOUNTING PRINCIPLES

Table 2 compares U.S. and non-U.S. accounting principles. The following discussion of Table 2 can be used as such or expanded upon, either way to be introduced throughout the intermediate and advanced accounting sequence as each particular principle is covered in the courses.

Asset Valuation

Historical cost is used as the asset valuation base in all of the countries considered in this paper—France, Japan, the Netherlands, Sweden, and the United States. However, a choice of historical cost or current replacement cost exists in the Netherlands. Indeed, the use of current replacement cost is becoming prevalent in the Netherlands (Al Hashim and Arpan 1988, 37), which does not really have accounting rules, but guidelines instead (i.e., there is no written code of mandatory standards in the Netherlands). The reason that replacement cost is often used in the Netherlands to value assets is that accountants in that country study current value accounting.

Asset Revaluation

While asset write-downs can be made in each of these countries, upward revaluations are prohibited in Japan and the United States, and are seldom done in France and Sweden. In the Netherlands the use of replacement cost accounting means that asset write-ups are common.

Consolidated Financial Statements

Consolidated financial statements are published in each of the five countries. In the United States, France, Japan, and Sweden, consolidation of nonhomogeneous subsidiaries is required. In the Netherlands, consolidation of nonhomogeneous subsidiaries is unacceptable.

While the equity method of accounting for unconsolidated investments is used in France, Japan, the Netherlands and the United States, in Sweden, the cost method is used for unconsolidated investments on consolidated financial statements. This is consistent with the Swedish emphasis on conservatism and the tax code. Conservatisim is also reflected in Japan, where the equity method is used in consolidated financial statements for unconsolidated subsidiaries; but the cost method is applied in Japan on the parent company's own financial statements. Pooling is allowed for use in business combinations in all of the countries considered in this study.

In each country, goodwill is reflected from business combinations. In France, while goodwill is recognized, it is permissible not to amortize it thereafter. The maximum period of amortization of goodwill varies in the other countries: 5 years in Japan and the Netherlands, 10 years in Sweden, and 40 years in the United States. It should be observed that goodwill from a business combination may be either capitalized or expensed initially in Japan, and either capitalized or immediately written off to retained earnings or income in the Netherlands.

Inflation Accounting

There are no inflation accounting requirements in the five countries under study. However, should historical cost be used in the Netherlands, then current cost supplemental data is highly recommended. Current cost accounting, which includes inflation adjustments, is allowed for external financial reporting in the Netherlands. In the United States, the Securities and Exchange Commission (SEC) requires firms affected significantly by inflation to disclose those effects in financial reports, although this is never done.

Inventories

Inventories may be reflected at the lower of cost or market in each of the five countries, though in Japan cost is the primary basis. Cost can be either FIFO, LIFO, or weighted average cost. LIFO can be used in the United States, Japan, and the Netherlands; but not in France or Sweden. LIFO is basically a tax avoidance method, and would thus not be acceptable in more socialistic countries like France and Sweden.

Generally, the same accounting and tax methods of inventory are applied in the United States, France, Japan, and Sweden. Regarding the lower of cost or market in France, market is defined as net realizable value. While the lower

Table 2. U.S. versus Non-U.S. Accounting Principles

	France	Japan	Netherlands	Sweden	United States
Historical Cost as the Asset Valuation Method	Required	Required	Allowed[1]	Required	Required
Asset Revaluation: Write-up	Allowed	Prohibited	Allowed[2]	Allowed	Prohibited
Write-Down	Allowed	Allowed[3]	Allowed[2]	Allowed	Allowed[4]
Business Combinations:					
Consolidated Statements	Required	Required	Required[5]	Required	Required
Equity Method	Required	Required[6]	Required[7]	Prohibited	Required
Purchase Method	Recommended	Recommended	Recommended	Recommended	Recommended
Pooling-of-Interests	Allowed	Allowed[8,9]	Allowed[9]	Allowed[9]	Allowed
Goodwill Amortized	Allowed[10]	Allowed[11]	Allowed[11]	Allowed[12]	Required[13]
Inflation Accounting	Allowed	Allowed	Recommended	Allowed	Recommended[14]
Inventories:					
Lower of Cost or Market	Required	Allowed	Allowed[15]	Required	Required
LIFO Inventory	Prohibited	Allowed	Allowed[16]	Prohibited	Allowed
Depreciation as Cost Allocation	Required	Required	Required	Required	Required
Research as Cost Allocation					
Capitalized and Amortized	Allowed	Allowed	Allowed	Allowed	Prohibited[17]
Interest Imputation on Long-term Receivables and Payables	Allowed	Allowed	Allowed	Allowed	Required

Notes:

1. There is a choice of historical cost of current value (replacement cost or net realizable value), though historical cost is the main practice.
2. Revaluation is allowed either in the financial statements proper or footnotes to reflect either current replacement cost, net realizable value, or present value.
3. Only permanent write-downs are made.
4. Permanent write-downs should be made. Write-downs of inventory and marketable equity securities to the lower of cost or market are required.
5. Consolidation of nonhomogenous subsidiaries is unacceptable.
6. In Japan, the equity method is not used on the parent company's own statements, which show investments at cost. Furthermore, the equity method is not used for investments having less than a 50% ownership.
7. The equity method is used for all long-term investments in common stock in the Netherlands.
8. In Japan, pooling is applied to mergers (A + B = A) and consolidations, (A + B = C), both of which are not common in contrast to acquisitions (A + B = A + B).
9. Pooling is not generally applied.
10. The principal practice is to recognize goodwill as an asset and subsequently amortize it.
11. The period of amortization is generally 5 years.
12. The period of amortization is a maximum of 10 years.
13. The period of amortization is a maximum of 40 years.
14. The FASB requirement is no longer in effect. The Securities and Exchange Commission requires firms affected significantly by inflation to disclose such impact in reports to the SEC.
15. While the lower of cost or market is the principal practice, inventory may be reflected at current replacement cost with the revaluation shown in a revaluation reserve.
16. LIFO is not often used, but if so it is applied to cost of sales, with the balance sheet reflecting current replacement cost.
17. Research and development expenditures that have alternative future uses are capitalized in the United States.

249

of cost or market is the principal inventory practice in the Netherlands, inventory may be valued there at current replacement cost.

Depreciation

Depreciation on property, plant, and equipment in all countries considered in this study constitutes a cost-allocation, not a valuation, process. In the United States there have been no standards issued by the FASB on depreciation accounting because the subject is not controversial. It should be noted that excessive conservatism in accounting for fixed assets prevails in Sweden and France, manifested in high depreciation charges.

Research and Development

Generally, research and development expenditures can be expensed in all five countries. However, there are exceptions to this rule. In the United States, expenditures with an alternative future use must be capitalized. However, in the other four countries, expenditures on development can be capitalized if technological feasibility of the product or process can be established and the costs can be measured. In the area of capitalization of R&D, non-U.S. countries tend to place more emphasis on professional judgment and give auditors more leeway concerning expected benefits of development costs. The U.S. practice of capitalizing only those costs with alternative future uses reflects the desire for uniformity in accounting for R&D as well as the fear of litigation when uncertain data are presented in a nonconservative manner.

Interest Imputation

While imputation of interest on long-term receivables and payables is GAAP in the United States, it is not the main practice in France, Japan, the Netherlands, and Sweden. In the United States there is a great deal of emphasis on "economic substance versus legal form" in financial accounting and reporting (Cairns 1990), probably the reason for the interest imputation requirement. In addition, the Financial Accounting Standards Board's Statements of Financial Accounting Concepts emphasize decision making over stewardship and imply the importance of present value by stating that accounting information serves to help predict the amounts, timing, and uncertainty of future cash flows. These concepts focus on economic substance over form.

To conclude this analysis of a small sample of comparative accounting principles, American accounting principles are generally more complex, more detailed, and less flexible than their counterparts in other countries. There is considerably more laissez-faire in accounting principles outside the United States.

ECONOMIC CONSEQUENCES

In attempting to explain nonculture-related reasons for differences in GAAP between countries, an instructor may wish to consider the economic consequences underlying specific principles, that is, the impact of accounting principles on the wealth and decision making of the various parties affected by financial reports such as preparers and users. The concept of economic consequences includes all economic ramifications associated with the principle in question, such as effects on the prices of the firm's securities, its cost of capital, debt covenants, management compensation plans, and potential regulatory commission actions against the firm.

Principle-setting bodies today are becoming much more aware of the possible economic and political consequences of proposed principles, and are using such consequences as a key factor in developing new accounting principles. Indeed, all decisions concerned with a change in accounting principle or the introduction of a new principle ought to have actual economic consequences, or else there is no justifiable reason for the change. In view of these consequences, compromises among alternative positions are clearly needed in the formulation of principles, so any attempt to associate pure logic with the formulation of accounting principles is bound to fail. By bringing the subject of economic consequences into the classroom, the instructor can emphasize the dynamic nature of setting accounting principles. Additionally, there is a trend in France, Japan, and Sweden to publish financial reports under U.S. GAAP. Volvo, for example, finds such harmonization to be desirable since the firm is listed on a number of international stock exchanges.

SUMMARY AND CONCLUDING POINTS

This study has reviewed selected accounting principles in five countries. We took eight principles and compared them across the five countries, and found differences in principles that can be explained, in part, in terms of cultural factors.

A comparative study of these principles should say something about accounting and culture: Are there significant disparities in accounting principles among countries? If so, why? To what extent do some countries follow other countries in developing accounting principles? Those questions have been the focus of this study. In view of all the factors that account for differences in accounting principles, harmonization of such principles on an international scale to any significant degree may seem unlikely. Harmonization may well occur among similar countries (e.g., in the EC), but probably not much beyond that. To the extent that a country increases its international business, harmonization will most likely be necessary at least in overseas capital markets. As long as the United States remains a leader in international commerce and accounting, it should exert influence over the nature of accounting standards adopted by countries worldwide.

One way of incorporating the comparative material presented in this paper into the upper-level accounting sequence would be to take a particular subject such as research and development costs and have the class consider and evaluate alternative principles of accounting for research and development. Subsequently, the instructor could ask the class which accounting principle would most likely be used in a particular country in light of the cultural factors prevailing in that country. Put another way, students might be asked to match accounting principles with specific countries and then to compare the principles they selected with those actually used in each country. This exercise should help students understand the rationale for adopting one accounting principle over another.

Another means of internationalizing the upper-level accounting courses is to have the class develop a conceptual framework, including the objectives and qualitative characteristics of financial reports (such as conservatism) for each of a set of different countries based on the cultural and economic conditions in each country. After formulating a conceptual framework for each country, students should be asked to apply that framework to particular accounting issues with an eye to development of specific accounting principles in each country. Subsequent to developing those principles, the students should be asked to compare and attempt to reconcile them with the actual principles in effect in each country under consideration.

Above all, the key question underlying this paper is: "Of what benefit is it for students to study accounting standards in different countries?" The answer is that such an examination shifts the perspective of the student from memorization to understanding our own GAAP. This point cannot be overemphasized.

One of the coauthors of this paper has applied the suggestions set forth in this paper in teaching intermediate accounting, and found that the emphasis on comparative accounting principles motivated all the students to consider the possible reasons for differences among specific GAAP in various countries. Moreover, the international students in these classes took a particular interest in the GAAP of their own countries and actually brought into class, for all the students to examine and comment on, financial reports issued in their own countries.

ACKNOWLEDGMENT

This study was supported in part by grants from the Wasmer Foundation in the School of Business at John Carroll University and the Arthur Andersen & Company Foundation.

NOTE

1. Let us briefly mention at this point the major differences, for example, between Anglo-Saxon and Continental European accounting: The latter in contrast to the former emphasizes the legal form of accounting principles, not the economic substance of financial reporting. On the other hand, Anglo-Saxon accounting is concerned with a "true and fair view," including, if necessary, consolidated financial statements.

REFERENCES

Accreditation Council Policies, Procedures, and Standards. 1986. American Assembly of Collegiate Schools of Business, St. Louis, Missouri.

Al Hashim, D.A., and J.S. Arpan. 1988. *International Dimensions of Accounting,* 2nd ed. Boston, MA: Kent.

American Accounting Association Committee on the Future Structure, Content, and Scope of Accounting Education. 1986. Future accounting education: Preparing for the expanding profession. *Issues in Accounting Education* (Spring): 168-95.

American Institute of Certified Public Accountants. 1988a. *The Accounting Profession in France.* New York: AICPA.

———. 1988b. *The Accounting Profession in Japan.* New York: AICPA.

———. 1988c. *The Accounting Profession in Sweden.* New York: AICPA.

Bloom, R., and M. Collins. 1988 U.K. and United States GAAP: A comparison. Unpublished paper, John Carroll University.

Cairns, D.H. 1990. Correspondence with the authors (February 7).

Gray, S.J. 1988. Towards a theory of cultural influence on the development of accounting systems internationally. *Abacus* 24(1).

Gray, S.J., L.G. Campbell, and J.C. Shaw. 1984. *International Financial Reporting.* New York: St. Martin's Press.

Hofstede, G. 1980. *Culture's Consequences.* San Mateo, CA: Sage.

Kroeber, A.L., and C. Kluckhohn. 1952. *Culture: A Critical Review of Concepts and Definitions.* Cambridge, MA: Peabody Museum.

Most, K.S. 1989. Chairperson's message. *Forum: International Accounting.* International Accounting Section, American Accounting Association (Spring).

Paton, W.A., and A.C. Littleton. 1985. *An Introduction to Corporate Accounting Standards.* American Accounting Association.

Porter, L.W., and L.E. McKibben. 1988. *Management Education and Development: Drift or Thrust into the 21st Century?* New York: McGraw-Hill.

Zeff, S.A. 1985. Some cultural speculations on the "forging" studies. Paper presented at the Workshop on Accounting and Culture, European Institute for Advanced Study in Management, Amsterdam, The Netherlands (June).

THE PERFORMANCE
OF MALE VERSUS FEMALE
ACCOUNTING STUDENTS:
EVIDENCE FROM THE THIRD WORLD

Rifaat Ahmed Abdel Karim and

Ali Mohamed Ibrahim

ABSTRACT

Several recent studies have investigated the effect of gender on students' performance in accounting courses. The results reported by these studies revealed that female students outperformed male students, although not all the results were statistically significant. The present study compared the performance of male and female students in an upper-division accounting course offered during the period 1985-1989 in a third world university. The findings indicated that the male students performed better than the female students. It is argued that differences in culture could be an important factor that limits the generalizability of the findings of the other studies and also helps to explain the differing results of those studies and the current one.

Advances in International Accounting,
Volume 5, pages 255-262.
Copyright © 1992 by JAI Press Inc.
All rights of reproduction in any form reserved.
ISBN: 1-55938-415-8

The issue of whether academic performance is related to student gender in the accounting classroom has been debated in a number of recent studies that have produced mixed results. The findings reported by Hanks and Shivaswamy (1985) and Fraser, Lytle, and Stolle (1978) revealed that female students did not significantly outperform male students, while the results disclosed by Mutchler, Turner, and Williams (1987) indicated that the superior performance by female students is statistically significant. However, the results of Mutchler et al. were not supported by Lipe (1989) and Tyson (1989).

According to Mutchler et al. (1987), the superior performance of female students could be attributed to four competing and complementary theories: (1) female students may be driven to perform better than male students in order to succeed in a stereotypically male profession, (2) female students are more success oriented and career motivated than male students, (3) female students may have a higher quantitative aptitude than male students, and (4) female students who major in accounting may be perceived by instructors as being more outstanding than males.

The research reported in this paper examines the effect of sex on the performance of students in an accounting classroom from a third world persepctive. This would help in determining whether the results reported in the study conducted by Mutchler et al. (1987) as well as the other studies mentioned above can be generalized to other countries.

The remainder of this paper is organized as follows. The next section describes the research design, and in the second section the research results are reported. In the penultimate section, the discussion of the results is presented and the concluding remarks are stated in the final section.

RESEARCH DESIGN

The data of this study were collected from a management accounting course which was taught by one male assistant professor for seven consecutive semesters during the period 1985-1989 at a university in one of the third world countries. The course is an upper-division course and is usually taken by students majoring in accounting as well as others who are majoring in other disciplines. Unlike some students in the latter group, none of the students majoring in accounting are offered courses in mathematics as part of their curriculum. Table 1 shows the percentage of the male and the female students who registered in this course in each semester. Table 2 gives the percentage of the male and the female students majoring in accounting and those majoring in other disciplines as well as the students in the latter group who had courses in mathematics and those who did not.

Both categories of students are required to complete four accounting courses before registering in this management accounting course. These courses are

Table 1. The Enrollment of Males and Females in Each Semester

Semester	Number	Percentage Enrolled	
		Male	*Female*
1	25	56.0	44.0
2	31	58.1	41.9
3	26	73.1	26.9
4	33	63.6	36.4
5	16	68.8	31.3
6	46	58.7	41.3
7	39	74.4	25.6
Total/Average	216	64.4	35.6

Table 2. The Percentage of Male and Female Students
Majoring in Accounting and Other Disciplines

			Other Disciplines					
Accounting Major			*Mathematics Background*			*Nonmathematics Background*		
No.	*M*	*F*	*No.*	*M*	*F*	*No.*	*M*	*F*
152	67.1	32.9	50	50	50	14	85.7	14.3

Notes: M = Male; F = Female.

"Principles of Financial Accounting," "Intermediate Accounting" (two courses), and "Cost Accounting." Students can choose the instructor with whom they would like to register because instructors are assigned to courses prior to student registration.

The same course outline and material were used in all semesters. The course grades were based on three objective examinations (i.e., multiple choice questions and exercises) which were graded by the assistant professor only. This helped to eliminate subjectivity in grading.

The same examinations were used in the seven semesters. The examination booklets were taken from students after each examination so that students in the following semester had no access to the questions. The use of the same examinations and grading policy in all semesters allowed us to compare student performance across semesters and to identify any performance pattern emerging during the period studied. All seven semesters were taught during the regular academic year (i.e., no summer sessions).

RESULTS

Table 3 shows the performance of both sexes across the seven-semester period. The table illustrates that for the whole class the male students performed

Table 3. Mean Course Scores for Male and Female Students

	Male		Female	
Semester	Mean	Standard Deviation	Mean	Standard Deviation
1	38.14	12.23	34.54	14.52
2	32.39	13.22	25.92	9.83
3	35.37	19.98	24.14	4.85
4	26.14	10.81	26.67	6.83
5	40.36	17.76	40.00	17.51
6	47.41	17.45	49.63	19.84
7	49.45	18.35	42.20	15.65
Average	39.54	17.96	35.99	16.94

slightly better than the female students in five out of the seven semesters. Nonetheless, in the two semesters in which the female students performed better than the male students, the differences in the means were very small. The male students had a mean score of 39.54 across the seven semesters while the female students had a corresponding mean score of 35.99. This difference was not statistically significant.

A number of analyses were performed to test for differences in the performance of the male students compared to the female students based on three independent variables, namely semester, major, and mathematics courses.

A two-way analysis of variance (ANOVA) was conducted to test the effect of the semester and sex on the students' performance. The results revealed that there was a statistically significant semester effect ($F = 9.69$; df $= 6$; $p = .0000$), and an insignificant sex effect ($F = 2.02$; df $= 1$; $p = .157$). The interaction term of these two variables was statistically insignificant ($F = .678$; df $= 6$; $p = .667$).

A two-way ANOVA also was performed to test the effect of the major on the students' scores. The results signified that there was no statistically significant major effect ($F = .156$; df $= 1$; $p = .693$). The interaction term between sex and major had a significant level of .073. This led us to further investigate the effect of sex among students majoring in accounting. The latter provided a more homogeneous group and helped in eliminating other variables like the effect of mathematics courses. The results indicated that the male students significantly outperformed the female students (t-value $= 2.17$; df $= 150$; $p = .032$).

In the nonaccounting major group, the female students performed slightly better than the male students, but not statistically significantly ($t = .71$; df $= 62$; $p = .483$). The mean scores of the female and the male students were 40.48 and 37.51, respectively.

A two-way ANOVA was conducted in order to examine if the students' performance depended on whether they had courses in college mathematics. The results of this analysis signified that there was no statistically significant mathematics effect ($F = 1.201$; df $= 1$; $p = .274$). However, the interaction effect between sex and mathematics was significant ($F = 4.825$; df $= 1$; $p = .029$). This finding led us to further investigate whether there was a difference in performance between the female and male students among the group which was offered courses in mathematics. It should be recalled that this included only students not majoring in accounting because, according to the present curriculum of this university, accounting majors do not take any mathematics courses. The results indicated that the female students performed better than the male students, but not statistically significantly ($t = 1.13$; df $= 48$; $p = .266$).

DISCUSSION

The aim of this study was to test whether the female students outperformed the male students in an upper-division management accounting course in a third world country. The results help to determine whether the findings reported in the study conducted by Mutchler et al. (1987), who found that female students significantly outperformed male students, can be generalized to countries other than the United States.

Three major findings were obtained that further contribute to this debatable issue. The first finding revealed that male students outperformed female students fairly consistently through seven academic semesters although the difference was not statistically significant. However, the second finding was that male students majoring in accounting significantly outperformed females in the same major.

These two findings contradict both the results reported by Mutchler et al. (1987) and those reached in the other studies, which found a trend of female students outperforming male students, although not always statistically significantly. A possible explanation for the differing results of the latter studies and the current study is that unlike the United States. where various professional fields are open to females, in many countries of the third world professional roles (including accounting) are still governed by traditions in the sense that many careers continue to be male-dominated spheres. Hence, given that the culture of a society tends to govern the role of sexes in the accounting profession, the females in the Mutchler et al. (1987) study being in the United States might have sensed a more challenging situation that called for fierce competition with males and thus a need to excel in achievement. However, this does not seem to be the case in most of the third world countries where high performance does not guarantee that females would have better opportunities in the accounting profession. Thus, ultimate rewards are not necessarily sensed by female accounting students.

The above analysis gains support from the findings reached by Lathan, Ostrowski, Pavlock, and Scott (1987) who surveyed a large number of graduating accounting seniors in the United States and concluded that female students expected to receive lower salaries, leave public accounting sooner, and become partners less often than males. If this is the level of motivation of female accounting graduates in a developed country like the United States, then it is expected that a similar group in a third world country would be far less motivated.

On the other hand, in the present study the males in the other disciplines failed to outperform the females of the same group. This could be due to the fact that females who are not majoring in accounting do not perceive the cultural constraints in their society that apply to the female accounting majors and hence, they (females in the former group) are expected to compete with males (i.e., males are not in an advantageous situation).

The third finding of this study was that the female students who had mathematics courses performed slightly better than the male students in the same group. However, the results were not statistically significant. This casts doubt on the theory advanced by Mutchler et al. (1987) claiming that the superior performance of female students in accounting courses could be attributed to their higher mathematical or quantitative aptitude. Two assumptions underlie this theory, namely (1) mathematics or quantitative aptitude ensures (or leads to) higher aptitude for accounting; and (2) female students who take upper-division accounting courses have higher mathematical or quantitative aptitude than male students (Lipe 1989). In addition to these two assumptions, there is a third (but an obvious) assumption which says that accounting is a discipline which requires inherent aptitude and this aptitude is directly related to performance.

This argument implies two hypotheses concerning students' performance in accounting courses, namely (1) students with mathematics background outperform those with no mathematics background, and (2) female students with mathematics background outperform male students with mathematics background. However, the findings obtained in the present study showed that neither of these two hypotheses is tenable. The students' performance was not found to be affected by whether or not they had courses in mathematics. In addition, although the female students with mathematics backgrounds outperformed the male students in the same group slightly (but not significantly), this might have been merely due to chance, as the nonsignificance implies.

The theme emerging from the above discussion of the three findings of the present study suggests that gender per se does not explain students' performance in accounting courses. Rather, the effect of gender may tend to be *contingent* on the cultural influence of society on the accounting profession. In this context, the effect of gender is treated as an intervening variable and the cultural influence of society is the independent variable. The implication of this argument is that the effect of gender on performance is not an outcome of inherent aptitude (e.g., mathematics and/or quantitative) abilities.

If certain societies are not yet receptive to the role of females in the accounting profession, this would limit the career opportunities of females in this field and hence, compared to male students, they might not be motivated to excel in studying for this profession, particularly at the upper-division level. On the other hand, the reverse is expected to be true in those societies which have started to provide females with career opportunities in accounting which until recently was a male-dominated profession.

CONCLUDING REMARKS

A number of studies have attempted to examine whether female students outperform male students in accounting courses. These studies have provided several competing and complementary theories to explain their contrasting results.

The present project studied the effect of student gender on performance in an upper-division management accounting course which was offered in a university in a third world country. Contrary to the findings of Mutchler et al. (1987) and the other studies quoted in this paper, the results obtained in the present project revealed that the male students outperform the female students fairly consistently. In addition, it was found that the male students majoring in accounting significantly outperform the female students in the same group. The results signified that neither mathematics had an effect on students' performance in accounting courses nor did the female students with mathematics background significantly outperform their male peers in the same group.

The differing results reported in the Mutchler et al. (1987) study and the current study is attributed to the effect of cultural influence in society on the career opportunities of females in the accounting profession. This implies that the effect of gender on students' performance in accounting courses is contingent on the influence of culture on accounting profession rather than on the influence of inherent characteristics of gender. It is along this line of thought that we believe further research in this issue should focus.

REFERENCES

Fraser, A.M., R. Lytle, and C.S. Stolle. 1978. Profile of female accounting majors: Academic performance and behavioral characteristics. *Woman CPA* (October): 18-21.

Hanks, G.F., and M. Shivaswamy. 1985. Academic performance in accounting: Is there a gender gap? *Journal of Business Education* (January): 154-156.

Lathan, M.H., Jr., B.A. Ostrowski, E.J. Pavlock, and R.A. Scott. 1987. Recruiting entry level staff: Gender differences. *The CPA Journal* (January): 30-42.

Lipe, M.G. 1989. Further evidence on the performance of female versus male accounting students. *Issues in Accounting Education* (Spring): 144-152.

Mutchler, J.F., J.H. Turner, and D.D. Williams. 1987. The performance of female versus male accounting students. *Issues in Accounting Education* (Spring): 103-111.

Tyson, T. 1989. Grade performance in introductory accounting courses: Why female students outperform males. *Issues in Accounting Education* (Spring): 153-160.

PART VI

INTERNATIONAL GOVERNMENTAL ACCOUNTING

GOVERNMENT ACCOUNTING AND AUDITING IN THE NETHERLANDS

Aad D. Bac

ABSTRACT

Government accounting and auditing in the Netherlands has improved considerably, at least as far as the provinces and the municipalities are concerned. National government is also gradually improving its accounting system and has recently introduced modified accrual accounting into the cash-based system. This paper attempts to define the differences between private and public accounting and to explain the reasons behind the differences, in the context of the Netherlands. Government auditing is done differently for the national government, on the one hand, and for provincial and municipal governments on the other. On the national level, auditing has been prescribed in a specific public way. On the lower government level the auditing function has been structured according to an accountability model and the analogy with the private sector is clearly visible. Nevertheless, government auditing is a specific professional field. A post-graduate course for government auditing exists and provides an opportunity for a necessary professional specialization. This paper also includes an analysis of the specific aspects of government auditing and argues that government auditing, being much older than private auditing, deserves to be respected more and given its own recognized profile.

Advances in International Accounting,
Volume 5, pages 265-288.
Copyright © 1992 by JAI Press Inc.
All rights of reproduction in any form reserved.
ISBN: 1-55938-415-8

INTRODUCTION

A recent volume in the so-called *Groene Reeks* (the "Green Series") published by the Association of the Netherlands' Municipalities (VNG, 1988) included a report by the *VNG-overleggroep Middeleninformatie* (VNG consultation group on resource information), entitled "Financial reporting as a policy-making instrument." Immediately after the introduction the booklet begins:

> In contrast to the situation in the seventies, many people working in the public sector nowadays consider the private sector to provide an example worth following. Senior civil servants style themselves "manager"; various parts of the organization speak of their "product" and their "business plan"; and where once the job satisfaction of the individual employee appeared to be the *raison d'être*, departments are now trying to become customer-orientated and are keen on internal contracts which reflect market forces.

This caricature of the situation was of course deliberate, but it does give a good idea of what has happened in fact.

Because of the pressure of dwindling financial resources, more attention had to be paid to cost control and efficiency, even for those processes of a somewhat abstract character within the public sector economy. In times of relative plenty it is easy to accept that this or that process "cannot be measured." In times of scarcity, however, one has to ask, "Is this absolutely true or is there perhaps something to be said for stricter control?" Scarcity forced us to return from euphoria to sobriety.

This new-found sobriety brought with it a return to the idea that the people in a department are a means to an end and not an end in themselves. Although some believe that "In the multitude of people is the king's honour" (Proverbs 14:28), subordinates have to be rewarded in order to keep them happy, so they must earn their keep. Fail in this and glory will crumble. Scarcity brought a timely reminder that the higher one climbs the harder one falls, and public employees started being careful. This did not mean, however, that they became a throw-away item; it is important to look after human resources, but there has been a shift of emphasis, an important one. Activities are no longer an end in themselves, and it is acknowledged that they should have a purpose. From now on government action will be less one-sided.

Where the critera used to be essentially legitimacy, justice and safety from prosecution, they now include goal orientation, effectiveness, and efficiency. The booklet from the *Groene Reeks* cited continues:

> This being the case, it is also understandable that the public sector is now paying greater attention to financial reporting, in the way that the private sector has had to do for a long time. Yet the function to the financial reporting of public sector institutions is completely different from the function of financial reporting in the business world.

I will now examine these differences, emphasizing public authority accounting and reporting in the Netherlands. I shall also attempt to place the Dutch situation in an international context.

MANAGEMENT AND ACCOUNTABILITY STRUCTURE

Public authorities differ in many ways from the business community. On the one hand, there are differences between the characteristics of the private sector and some of those of government. On the other hand, public authorities are also active in fields in which the private sector is or could be active. These two areas of activity are usually described as the government sector and the business sector (of government). This distinction is also found in the regulations governing such activities. At central government level, there is a Government Enterprises Law in addition to an Accountability Law. The provincial and municipal accounting regulations both have separate sections relating to enterprise activities. There are different reporting standards for central government enterprises and for similar activities carried on by provinces or municipalities. What strikes one immediately is that, at the central government and provincial level, the crucial question is whether there has been an official registration as a business enterprise, whereas at the municipal level it no longer matters whether a separate business has been established (in contrast to the situation prior to 1985), but depends instead on whether the activity can be seen as having the nature of a business. If so, then the different reporting standards apply, regardless of the type of business and the chosen form of legal entity.

Because we can identify two different sectors, it also makes sense to consider the differences for the two sectors separately. One thing which the two sectors have in common, however, is their executive structure. We are used to seeing an organization in the form of a pyramid. At the top we have the senior management and beneath that a number of levels, depending on the organization concerned. This scheme can also be applied to public sector organizations, but in this case the picture is incomplete, because there is also an executive organization which is more or less divorced from the bureaucracy. The executive can be represented by an inverted pyramid. On top there are the representatives of the people, then below them an executive board, and at the bottom, the individual office holder. This is the level at which the politicians and the administrators meet. The different cultures of their two organizations are reflected in the way political and commercial, financial and intangible aspects are balanced in a complementary management organization.

Political and intangible aspects will probably be more important to the executive organization, and commercial and financial aspects will be more important to the bureaucrat. This conflict may not always be very clearly visible but the decision-making process should be reflected in the accounting and reporting systems and especially in the explanatory notes to the accounts.

The complementary nature of the management is a factor complicating the decision-making process. This concerns the individual member of the executive board and his or her relation with the senior administrative officers. In this regard, there are also differences between central government and the other public authorities. Where in central government there is individual ministerial responsibility, the provinces and municipalities operate with collegial executives. This means that ministers have to defend their own annual accounts before the parliament—at least, that has recently been the case. The central government accounts are based on an act governing the collection and transfer of funds, and are prepared by the Minister of Finance, if one can describe as accounts, let alone annual accounts, what is merely an annex attached to the act finalizing the budget. I will return to this subject later.

At provincial and municipal levels, it is the boards composed of members of the provincial executive and the mayor and aldermen, respectively, who are collectively responsible for control. The control they exercise is based on single blanket accounts. Apart from that, these groups are also only temporarily left to their own devices, because, owing to the monistic nature of the control which is exercised, the Provincial Council and the Municipal Council more or less take over the accounts from the Executive Board. The members of the board, except for their chairman, are after all also members of the Provincial Council or the Municipal Council. All members of the Provincial Council or the Municipal Council in turn render joint external accounts of their policy and the execution thereof.

At the central government level, the control exercised is essentially dualistic. The executive and legislative (parliament) are explicitly separate. Parliament has no part in day-to-day government. The executive, therefore, renders accounts of both policy making and its execution (i.e., control of the policy). Parliament contents itself with a supervisory or watchdog role.

REPORTING BY GOVERNMENT ENTITIES

I have referred to the minor role played by annual accounts as far as central government is concerned, because an annex attached to the act finalizing the budget is all there is. The rendering of account is very much an afterthought: the budget is the main issue. The government proceeds from budget to budget, with spring and autumn statements by the finance minister in between, but these are actually only repeats of the original budget with minor course corrections. Long before any accounts are drawn up, a new budget has already been presented. Until recently the accounts were about seven years late. To avoid trouble, however, provisional figures were produced fairly promptly—

figures which were good enough to serve as explanatory information for the next budget. Then everything could go on as usual and the figures could be dealt with later, but they were literally "of no further interest." Even if provisional figures do not differ from the final ones, which for various reasons can occasionally be delayed even longer, it does matter. If figures are used for obtaining resources for a specific government task, they serve a totally different purpose from that of figures presented for rendering account with respect to that task. In the latter function they are supposed to give an insight into the degree to which the intended task has been achieved. To perform this function they need to be accompanied by an explanation. Accounts of this nature should be subject to audit. An important criterion of this examination is whether the information given, and especially that given by the accompanying notes, is adequate. We must admit that the government makes far more information public than the private sector does, especially with regard to budget figures. This cannot, however, make up for missing out one whole step, rendering account as a part of the budget cycle. Submitting period accounts must be accorded full value, and not be a mere formality. As far as central government is concerned, there is accordingly still much that is missing.

REPORTING IN THE GOVERNMENT SECTOR

I shall now compare this picture with the situation in the business world. Because the government sector is more specific than the business sector, I shall examine this sector first. The characteristic feature of this sector is spending income. The main problem, when it comes to use of resources, is shortage of funds. The various possible uses must compete with each other. These problems of allocation and choice mean that the budget in the form of accounts with a zero balance between income and expenditure is ideal. This appears to optimize the use of available resources. The critical success factor in this sector is therefore definitely not found in the balance of the account. Success is measured in terms of the various ways of spending the resources, not the spending as such, but as determined according to the benefit gained.

Functional Accounting

The traditional functional grouping in the annual accounts is therefore logical. The remarkable thing is that, beginning with the 1989 budget year, central government will also be grouping the figures according to areas of policy, although the items will only be the so-called program costs. There will be two cost elements for each departmental budget, covering the entire cost of the administrative apparatus, one for personnel and the other for materials. From the reporting point of view, this is not an improvement. In financial terms, we cannot see how the program costs relate to the costs of the apparatus or, for that matter, to the product of that apparatus.

It should be emphasized that the municipal functional accounting system is designed to give an insight into the input to the areas of policy which the government is addressing. This is a reporting model that is suited to rendering account of policy. It is, therefore, only reasonable to ask whether it is the most suitable form for calling the executive to account as well. It has been suggested that the executive accounts should have an organizational grouping, or more precisely, should reflect the structure of the municipal economy. It is interesting to note that this is precisely the situation which used to obtain with respect to central government. There is much to be said for making the official in charge accountable to the executive board by means of a reporting system that reflects the organizational structure as it exists. This would enable results to be closely identified with responsibilities. The question then arises as to whether this should also be done with regard to the accountability of the individual office holder in the board. The problem in this case would be that, in view of the collegiate nature of the board, constitutionally there is no such thing as accountability of the individual office holder. The next question is whether the accounts of the board should be drawn up from an organizational angle. This is an interesting thought, because we would then finally discover what the difference ought to be between the executive accounts and the policy accounts. These two sets of accounts would, of course, have to articulate. In this way the members of the board would receive both organizational accounts, serving to discharge them of further responsibility for their administration, and functional accounts in anticipation of the need for external reporting by the Provincial Council or the Municipal Council, of which they also are members. It may seem quite a lot of work, but the idea deserves further consideration.

Gross Accounts

The process of making choices always leads to divisions, given the limited resources available at the time of allocation. If extra resources become available, it is of course for the representatives of the people to decide—at the instigation of the cabinet or board—which uses should be made of additional funds, or to what new purposes funds economized should be put. This right to decide budget allocations is so important that it deserves protection. This is reflected in the requirement to produce a gross account. If in a particular field of activities a certain amount of income is included which is generated by the activity itself, then growth in such income must not automatically lead to a corresponding growth in expenditure. The unexpected extra income or lower expenditure should be re-allocated in subsequent budget rounds just as any contrary results would be. The gross accounting for the allocated amount is accordingly an important element in the rendering of account. That is why the budgeted amounts are also stated in the annual accounts of a government organization. There must also be some tangible evidence of safeguarding the right to draw up budgets. This can be achieved by strict credit control, which means supervision from an early stage so as to ensure that the expenditure can be met both qualitatively and quantitatively out of the sources earmarked for the purpose in question. The budgeted amounts are regularly adjusted. This is logical in view of the constant changes taking place in society. The reasons for budget adjustments can be:

1. additional new policy,
2. changes to existing policy,
3. changes to take into account overspending, and
4. budget finalization.

According to formal rules, the budget has to be finalized by a given date. Adjustments to the budget which are not made in time cannot be included in the annual accounts. From this we can conclude that changes which are made in time can or indeed must be taken into account. This means that annual accounts show the same amounts in both the budget column and the actual column. As I see it, therefore, the effort of producing the figures is out of all proportion to their information value. All the figures tell us that the budget was adjusted in time. If a decision on the adjustment of the budgeted amounts is to be made on meaningful grounds, then the discussion on this adjustment must be seen as a forerunner of the discussion of the annual accounts and, moreover, as a discussion outside the context of the annual accounts. In this respect the finalization of the budget undoubtedly constitutes an unintentional weakening of the process of discharging the responsibilities of the executive in dealing with the annual accounts. Moreover, separate treatment of the budget finalization is not necessary. A properly drafted set of notes to the annual accounts could provide the necessary insight into the relationship between budgeted and actual expenditure. The analysis of differences between "original and additionally approved policy" and realization gives insight into how much the budget has been overspent and, if everything is as it should be, the reason for the overspending. Implicit in the adoption of the accounts is retroactive approval of the finalization of the budget in the first place.

Stabilizers

Budgets are the authorized outcome of the allocation process. This authorization process not only gives content to the allocation, it also serves as a substitute management tool. Such a substitute management tool is necessary because government activities in the strict sense are not carried on to serve a market. In the government sector, there is hardly ever a market in the traditional sense. The market mechanism performs a comparison of the various suppliers and the various customers, and brings them together in equilibrium. This leads to a price that reflects cost to the suppliers, and the decision to buy or not to buy decides whether the suppliers stay in business.

There is, of course, a "market" in the creation of benefit, but only in an abstract sense. In this "market," negotiations take place through elections, lobbying, public debate, and even demagogy. No price is stated, so there is no mirror to reflect the cost, nor is there a mechanism which will signal a possible surplus of supply over demand. In other words, this market lacks a number of built-in stabilizers. It urgently needs these stabilizers, however, so suitable facilities need to be provided. The authorization procedures as such are not really suitable, inasmuch as the formalities are inadequate. An insight needs to be given into the extent to which the allocated resources have been usefully applied, that is, the extent to which they have served their purpose and have not been relatively too high. This means some kind of measurement of performance.

Performance Monitoring

Performance monitoring must also be reflected in the notes to the annual accounts. This could take various forms. As far as the financial aspects are concerned, performance monitoring can only take place with reference to the budget, provided the critical content of the budget is high enough. Public authority budgets, however, are not noted for giving precise terms of reference. Regardless of the fact that there is a tendency to bump up figures in order not to have to be constantly asking for additional allocations, there are several more objective reasons for the low critical content. The long period of preparation is one of them, but far more important are:

- the above-mentioned lack of a market,
- the often abstract nature of the service performed,
- the varying intervals between cause and effect (sometimes effects will only be seen over a longer term), and
- the occasionally multiple interdependent effects.

It is safe to say that in many cases it is difficult if not impossible to adopt standards. This means that material for comparison purposes must be acquired in other ways. Where possible, therefore, the following techniques are employed:

- external budget comparison,
- internal budget comparison,
- cost analysis,

- measuring or estimating of effects, and
- measuring of effort.

Operational and Capital Expenditure

There is another difference which is worth mentioning, that is, the existence or otherwise of separate treatment for capital expenditure. The difference between what is generally referred to as running expenditure and capital expenditure is not made everywhere. Indeed, this difference is not made at all by central government. The most serious objection to this is that it is a violation of the principle of matching. Especially when we think of the enormous sums of money handled by the government in the Netherlands, it is wrong not to have an insight into the size of a benefit which is supposed to match a given expenditure for a given period. Various criteria are available for differentiating between the running expenditure, that is, costs and charges, and capital expenditure, that is, investment. Without going too deeply into the subject we could mention:

- the criterion of usefulness,
- the criterion of periodicity, and
- the criterion of capital.

The same, of course, also applies to revenues. As long as the overall situation is not ideal in every respect, failure to make such a distinction will over- or under-endow, future generations and the people should be informed in particular about the latter. If we do not have separate treatment of capital expenditure and a full balance sheet, it is impossible to see how well or badly we are acting with respect to posterity. I therefore consider the lack of a separate treatment of capital expenditure to be a serious omission.

COMPARISON OF PUBLIC ENTERPRISES WITH THE PRIVATE SECTOR

Public sector enterprises, belonging to the business sector of government, are different from the private sector but the differences are hidden in the similarities. The similarities lie in the fact that public sector enterprises do actually operate in markets and usually they have a clear, generally straightforward purpose. In addition, their budgets take on more the form of an economic plan. The critical success factor with this kind of enterprise appears to be the profit. All of this is only partially true, of course.

Market Intervention Purpose

The government never operates in a market by accident. There can be no doubt about the consciousness of the decision to embark on a particular activity in the first place, but it may be some time before people realize that the original reason why the government decided it needed to perform a given task no longer exists. The motives for government enterprise are well-known. They are not based on the usual economic driving forces that prompt the entrepreneur to enter a market. The government always has another and perhaps more compelling reason for entering a market, usually to do with intervention. The public utility companies are a good example of this, because their products are counted as necessities of life. Governments were, therefore, not prepared to accept the threatening monopoly position of the suppliers, due to the nature of the distribution systems, to be left to free market forces. Another example can be seen in public transport. In this market there would be no suppliers at all at fares that are considered reasonable, so the government is obliged to take action to ensure the continuity of public transport services.

As a consequence, the policy underlying market intervention needs to be translated into a target result or, if you will, an acceptable result. The discrepancy between a negative target result and a normal return on capital employed needs to be covered in some way or other. In this context, it does not make sense to speak of a negative result of this kind as a loss. The contribution from the public purse which is necessary for this essentially unprofitable operation to continue should be looked upon as a revenue. Treating it as such will produce a profit. Where the government subsidy consists in just wiping out the negative balance, the approach referred to is not applicable. With this kind of open-ended financing, there is no possibility of controlling the amount of the government subsidies. That is also true as far as the explanation of the underlying policy is concerned. Adequate reporting ought to provide an insight in the aforementioned manner.

The nature of the activities carried on by government enterprises differs from the activities which we normally associate with the tasks of government in the strict sense. That means that a different significance should be attached to the bottom line of the accounts of government enterprises. Income and expenditure in the case of these enterprises are interrelated. In order to generate more income, additional expenditure may be unavoidable. It is not a question of allocation. We are after all generally concerned with but a single purpose. There really is no need for a separate authorization of such additional spending. The budget in this case assumes much more the nature of an economic plan, a model of the enterprise. It is therefore understandable that government enterprises which are forced to don the mantle of private legal entities regularly omit budget figures from their accounts. Government enterprises which are organized as statutory bodies are formally prohibited from doing so. In my view, one can draw one's own conclusions from this restriction.

THE BALANCE SHEET

A balance sheet is by no means commonplace in public sector accounts and, indeed, is a fairly recent phenomenon, at least as far as the Netherlands is concerned. The public sector has at the national level been producing "trial balances," but they are statements which are separate from the accounts as such and can make absolutely no claim to being comprehensive. The question of what a balance sheet shows has different schools of thought. The so-called monists argue that a balance sheet can only fulfill one function at a time: either *static*— a snapshot of the financial position at a particular moment in time; or *dynamic*— a statement of the financial position at the transition between two periods, drawn up with the object of determining the result for the elapsed period. The dualists, on the other hand, maintain that the balance sheet can perform more than one function. Limperg, with his replacement cost theory, can be seen as a dualist. If by capital we are to understand the value of the business entity, then, according to business economics, that means the present value of projected future earnings. This may be expressed by the discounted value of future cash flows. Owing to the uncertainties of the future, it is difficult if not impossible to determine this value. Even if it were possible, prudence would dictate that subjective goodwill should be excluded from the presentation.

But how do we treat the value of an economic entity based on a totally different economic philosophy, a philisophy not geared to profit but to utilizing the income obtained from taxation for maximum benefit? There is no possibility of present value here. As far as the balance sheet is concerned, therefore, the value of an economic entity cannot be turned into a suitably practicable concept. The draftsmen of the Netherlands Civil Code appear to have struggled with the same problem. This is apparent from Section 362 of Book 2 and the note on this section. The section stipulates that the balance sheet must give a true and fair view of the "capital." In explaining what is meant by capital, the note states that this should not be taken to mean the value of the enterprise. It is far more a question of the financial position, which must be accurately reflected within the scope provided by the adopted valuation principles. By allowing for the adoption of different valuation principles, the legislators imply that a true and fair view of the financial position can be given by different means.

Converting to a Balance Sheet

Obviously a newly drawn up balance sheet will show signs of its heritage. There will have been no time to formulate a vision of what the balance sheet represents. Initially, the balance sheet will be no more than a closing statement. The Civil Code may state that the assets and liabilities shown do not have to be any more than a reflection of the financial position as determined according to the adopted valuation policies, but, even on this basis, the majority of the balance sheets produced by provincial and municipal authorities do not yet begin to give a true presentation of assets and liabilities. Before this can be achieved, a solution will first have to be found to the whole problem of valuation with respect to public sector economic activities. In other words, appropriate valuation principles need to be found. These will have to take into account legacies from the past in the form of accumulated depreciation to date, investments that have been written off to expense, and infrastructure financed by land sales. Such practices result in low if not totally eliminated book values. Such treatment not only conflicts with a consistent valuation policy but is also detrimental to a balanced allocation function. It also creates problems, should the assets concerned ever need replacing. The replacement costs would have to be funded in full and thus form a drain on the available finances.

A EUROPEAN COMPARISON

A study report by the Council of Europe (1985), has this to say on the subject:

> The national contributions demonstrate the variety of the solutions adopted by members. One of the main reasons for this variety is the close connection between the institutional structure and the sometimes dogged attachment to a particular theory. However, there are also convergent factors, for instance with regard to objectives.
>
> Rising indebtedness and limited fiscal resources necessitate increasing resort to financing by users (the profit principle one might say). The growing importance of thinking in terms of costs is a common characteristic of all countries, and this has considerably influenced the development and, to a lesser degree, the convergence of systems. The repercussions of this development upon local and regional authority accounting are very important. They first require a distinction to be made between current and capital transactions. In addition, a balance sheet of assets must be drawn up, which is not incorporated in the public accounts in the strict sense but is presented in the form of a separate file. The degree of accuracy, the methods of calculation and the definition of costs vary not only from one country to another but sometimes even between different branches of the same authority.

A comparison between our more or less direct neighbors, Belgium, Luxemburg, France, the Federal Republic of Germany, and the United Kingdom, reveals the following with regard to local government.

There is considerable similarity in regard to the distinction between services involving running expenditure and capital expenditure. The explanation for this can presumably be found in the rule that forbids running expenditure from being financed other than out of current income—the "golden rule of finance." Only in the United Kingdom is this not laid down officially because the autonomous position of the local authorities under U.K. laws does not admit the imposition of rules by central government. Here, the professional body, the Chartered Institute of Public Finance and Accountancy, plays a harmonizing role. In all the countries concerned, there is a primary functional subdivision with a secondary analysis according to economic categories. In the sole case of Luxemburg, I was unable to ascertain whether this subdivision was prescribed by law.

There is greater diversity in the field of allocation methods. The Federal Republic of Germany, France, and Luxemburg apply either a simple cash or modified accrual system. Belgium also introduces estimates, and that also applies to the United Kingdom, so that the systems of these countries can be considered as income and expenditure systems.

Apart from Luxemburg, and in my view, also the Federal Republic of Germany, a balance sheet constitutes part of the annual accounts. In Belgium, this will now include fixed assets. Moreover, in the past, this balance sheet was not prepared as an integral part of the accounting system. That the full implications of the growing interest in financial management have not yet been realized is illustrated by the fact that not all the countries make a distinction between depreciation and repayment (i.e., Belgium, Luxemburg, and the United Kingdom). That is also the case with respect to allocation of indirect costs and internal charges. I get the impression that this certainly does not occur in Luxemburg and the United Kingdom, and presumably not in the Federal Republic of Germany either, although it does in Belgium to some extent and certainly does in France. All in all, the Netherlands does not come out of this comparison too badly.

The above comparison concerned mainly local authorities. Regional authorities generally have a system that corresponds either with that of central government or with that of local authorities. They therefore provide less interesting material for comparison. Finally, therefore, I should just like to touch on one other point regarding the situation as it applies to government in the Netherlands.

The Netherlands is one of those countries with a legal tradition based on the Napoleonic Code. It is hardly surprising, therefore, that we have a fairly uniform and structured system. That ties in with the conclusions drawn by Lüder (1989). However, there is a big difference between central government, and the other tiers of government. In local government in particular in the Netherlands, there are signs of a cautious movement away from actual legislation toward the laying-down of quality requirements.

THE EMANCIPATION OF GOVERNMENT AUDITING

Government auditing has received substantial attention in the Netherlands as a special field of professional interest after the introduction of a post-graduate course in 1988. This evolution coincides with a growing interest of politicians, including administrators, in audit reports. This interest is to an important degree the consequence of some spectacular cases of administrative misbehavior. Another reason is the need to freeze or even cut expenses. Stevers (1979) found that the courts of audit in the various European countries evolved from councils of noblemen, prelates, and other personal advisors of monarchs. The medieval monarchs used such councils to support them in the areas of legislation, justice, and administration. In those days there was no division of powers. Stevers shows us that the position of courts of audit has developed in strong relation to political changes. In the cases of absolute rulers, audit institutions worked for them; in more democratic societies, for the parliament.

These conclusions are supported by findings about civilizations in the even more distant past. The Shanghai University of Finance and Economics and the Center for International Accounting Development of the University of Texas at Dallas have undertaken a combined study project into accounting and auditing in the Peoples Republic of China. From this 1987 study it appears that about 500 BC a government audit became visible in instructions to all levels of local administration to report to the royal court on:

- the change of population,
- the increase and decrease of cultivated land, and
- the receipts and disbursements of money and grain.

From the fact that demographic information was to be reported it can be concluded that people in ancient times clearly understood that labour is a capital good. This is a nice example, in which it took centuries for an idea to be revived; human resource accounting "in statu nascendi."

This "hearing" was extended in the following period. During the successive dynasties, local officials were required to report to the royal court about the examiniation of such statements of income and payments of revenue, which were already laid down in books.

At about the same time occurred the so-called Babylonian exile of the Jewish people. The 6th chapter of the Book of Daniël in the Holy Bible shows that Daniël was charged with an auditing assignment. The passage reads as follows:

> It pleased Darius to set over the kingdom a hundred and twenty princes which should be over the whole kingdom;
> And over these three presidents of whom Daniël was first: that the princes might give accounts unto them, and the king should have no damage.

Stevers demonstrated that courts of audit are increasingly directed to parliament as democratic governments develop.

Present Situation

If we are to compare Stevers' conclusions with the present situation, we must not forget that the court of audit is a very specific example of an auditing institution in government. The constitutional position and the tasks of the Dutch Court of Audit are so special that the law on registered accountants shows great respect and cautiousness when dealing either with the court as such or with the registered accountants employed by the court. Also, even if democratic administrations exist many variations are possible. In a decentralized unitary state, as is the case in the Netherlands, the different governmental levels have a certain coherent autonomy. That is why it was necessary to develop an adapted accountability structure. The position of the court of audit is to an important extent an effect of the national level being the "highest" in our type of government. Parliament has a generally supervisory role in its relation to government. To ensure that the contribution of the highest auditing institution to the task of parliament is as effective as possible, a broad assignment is of great importance. As parliament combines its supervisory role with a role as co-legislator it is proper for the auditing institution as a constitutional institution to be also independent from parliament.

When looking at the provinces and the municipalities another balance between accountability, supervision, and audit can be seen. It is remarkable that the model which exists there is more of an accountability type than on the national level. The accountable provincial or municipal administrator is required to authenticate his accounts by an auditor's report, both in a short and a long form. This set of documents also serves a double purpose: first that of discharge by its own "parliament"—the provincial or the municipal council; and second, in a process of supervision of the budget by the "superior" government, the Ministry of the Interior for the provinces, and the provincial executive for the municipalities. This is essentially different from the situation on the national level, where the accountable minister is not required to present an auditor's report with his accounts. Therefore, parliament communicates directly with the constitutionally independent Court of Audit. This relation is more an inspection than an accountability supporting audit.

The accounts of an accountable minister are subsequently submitted to the court of audit, which is not limited to the accounts as such. It also looks at the financial management. However, it results only in a discharge procedure; there is no separate process of supervision, as is the case for provinces and municipalities.

IS THE SPECIALTY OF GOVERNMENT
AUDITING PROPERLY UNDERSTOOD?

Government accountancy and its separate audit function are not new. A recent book on the history of accountancy in the Netherlands starts as follows (de Vries 1985):

> The profession of accountants, as it is exercised nowadays, finds its roots in the last century, when economic life in the western world, on a new foundation, experienced a period of fast and strong growth.

The author does not look further back, but he comes to the conclusion that in ancient times financial management and accounting were of a very primitive character. De Vries states:

> In such mostly traditional agricultural societies the thinking of managers is not governed by economic points of view. Antiquity, too, saw the pursuit of gain and wealth, but this was not in the first place directed at the economy as an essential and inexhaustible source of income and subsequently of spending. In general in ancient times wealth is the reward for political, military and religious power or status and not for economic activity.

In this respect however, it was logical that a distinct audit emerged in the government environment. In the totalitarian regimes which then were normal, the development of complex organizational structures was seen in government long before the private sector showed a comparable development. Therefore, a need for a coherent set of specific auditing procedures arose in government in ancient times. In the private sector such a situation did not become common practice before the industrial revolution, with its increase of scale and the division of ownership and management.

A Misunderstanding

> During the first 5000 years of primitive accountability the auditor as an independent expert on business-economics, on administrative organization and management-information and on auditing of business-economic information constitutes an anachronism, but after the emergance of double-entry-bookkeeping and during the centuries thereafter, he was someone waiting in the wings. Not before the division between ownership and management at the end of the nineteenth century would the gates open (de Vries 1985).

This statement is both true and not true at the same time. If a definition is used which relates to circumstances in the twentieth century, it is logical that such a phenomenon is not found in other periods with completely different characteristics. In addition the definition only refers to the private sector. If we look at society as a whole the definition does not hold. What is wrong? In the first place, there is a one-sided fixation on the private sector (see above). In the second place, there is an overestimation of the influence of double-entry bookkeeping.

De Vries's framework was the description in the *Encyclopedie van de Bedrijfsceconomie* (1970). This may have put him on the wrong track. In the public sector of the Netherlands, double-entry bookkeeping was recently introduced in provinces and municipalities. At the national government level this has not yet happened. Is auditing in government, in light of de Vries's statements, *unnecessary, impossible,* or *more difficult?* More difficult it may be, but to say that it is unnecessary or impossible is to talk nonsense.

- The need of audits of governments and governmental agencies is shown clearly by the different ways in which the vox populi makes itself heard. So, unnecessariness is out.
- The impossibility is refuted by the age-long existence of auditing institutions. For 75 years now accountancy firms have been active in this field, if we may take VB Accountants as an example.

Specific Characteristics

But the audit in a governmental environment proceeds in a different direction. How could it be otherwise, when the economic processes go in a different direction? The cost and the income of governmental institutions are not causally related in a sense that cost does not precede income, as is the case in an institution which produces for the market. It is just the other way around: a government first imposes taxes on its subjects and afterwards decides what will be done with the available money. The situation is characterized by choice or allocation in an income-distributing institution in which spending is not always coupled with exchange. Not only is producing not always market-directed, often a market does not exist at all. That is fundamentally different from the situation in income-acquiring households. A system of authorization is like a checking instrument that replaces the market mechanism and its inherent possibilities of internal control. The importance of accounting for capital therefore has another relevance in a governmental environment. The absence of double-entry bookkeeping as such definitely does not need to be indicative of a low developed level of accountability, as de Vries claims. The need for audit, in conformity of course with the state of the art in each period, necessarily had to arise in governmental systems.

IS GOVERNMENT AUDITING FUNDAMENTALLY DIFFERENT?

A totally different question is whether government auditing is fundamentally different from private sector auditing. There are differences in detail that indeed are not unimportant, although they do not lead to a totally different profession.

I would distinguish between two main groups of differences in detail and give some examples of both. These groups are (1) environmental factors and (2) task inherent factors.

Environmental Factors

A first example of differences in detail originating from the government environment can be found in the importance of lawful acts by and in governmental organizations. The legislator more than anyone else has to conform to the law, or otherwise the citizens could no longer expect legal security and justice. This has the effect that the environment of the audit of government organizations is dominated by a large number of rules and regulations. For reasons of justice this system of legislation is complemented with unwritten law. This unwritten law is described as *common law*, and also as *principles of proper administration*.

The fact that practice is in compliance with regulations has an important influence on the sociopsychological characteristics of the official organization, with striking consequences for the organizational culture. The shaping and the operation of the organization are affected. A useful audit function must grow out of an adequate knowledge of these characteristics of an official organization and their implications.

This is also the case with respect to a second important factor that affects the culture of the official organization, namely the continuous interaction with the political organization. Official organizations are managed by a structure of complementary management. The actors in this complementary management are on the one hand the administrators and, on the other, the executive of the political organization. In the decision-making processes, in which the administrator participates, in order to prepare decisions a weighing of criteria takes place. We could say that this is the arena in which the different rationalisms which Snellen (1987) distinguishes fight for dominance, until a (possibly temporary) balance in the co-existence of these rationalisms arises. Several adventitious circumstances may give a particular rationalism the opportunity to dominate disproportionately. We are sometimes tempted to call that mismanagement, although Snellen is kinder in this respect.

On the basis of his rationalism theory Snellen makes an interesting comparative analysis of public administration, business administration, and management in nonprofit organizations. In public administration four rationalisms interact: (1) political, (2) juridical, (3) technico-scientific, and (4) economic.

In decision-making processes within nonprofit organizations the political and the juridical rationalism play a more moderate role. In business administration the role of the political and the juridical rationalism is hardly recognizable. In this sequence these two rationalisms gradually change from decision criteria to existing conditions. The level of complexity of the decision-making processes and the degree of difficulty of an effective and efficient policy and management strikingly coincide with the complexity of the rationalism structure. The advisory participation in these decision-making processes and the undergoing of their outcomes has an influence on the behavior of officials which should not be underestimated. The repercussions on the culture and the working methods of the organization are evident here too.

In the foregoing attention was given to the effects on the government environment of the political and the juridical rationalisms. We can learn from Snellen that the technicoscientific and economic rationalisms are of a more general character. However, economic rationalism in government is not the same as in the private sector. Apart from the reverse flow of resources and the implications thereof on the management structure of governmental organizations, especially in national government, microeconomic and macroeconomic theories often conflict. In government the economic rationalism thus has a more or less schizophrenic character.

Task Inherent Factors

A first important point concerns the position of the accountant in government. On the Dutch national level external audit exists (by the court of audit), but this is not external audit by an accountant in the usual sense. The departmental audit institutions are established in order to support departmental management and primarily have an internal audit function. The repercussions thereof can be found in the Resolution on the task of departmental auditing services. The audit of the annual accounts is only one of the elements of the audit task, whereas an audit is usualy perceived as accountability oriented. It adresses the question: How adequate is the view offered by the figures presented in relation to the information need of different users of the annual accounts?

The task of the departmental audit services contains several elements which are instrument-oriented or oriented toward the content of the information offered. Instrument-oriented is the mandate to audit the management accounting as such. Mutatis mutandis, this is also the case for the instruction separately to evaluate the system of internal control which is included in the administrative organization. Content-oriented is the task element that concerns the financial management. Although this is limited to the degree in which it appears in the accounts, it is still strictly not accountability directed.

Genuine accountability-oriented task elements do of course exist too. The accounts should not only be audited in order to evaluate their contribution to an adequate management, but also because of the accountability of management which must emerge from the annual accounts. Thus the audit of the accounts anticipates the accountability-oriented task of auditing the annual accounts.

In Dutch provinces and municipalities too, financial management and bookkeeping as such are substantial objects of audit. This can be derived from the provisions of the provincial law and the municipal law which regulate the assignment of the accountants. That the outcome of the audit should also contain a report on the soundness of the annual accounts is provided for by other articles in these laws.

The content orientedness and the accountability orientedness thus appear in lower government too. So far this is not yet the case with the instrument orientedness. In my opinion there are two explanations for this. In the first place these provisions in the provincial law and in the municipal law stem from a period in which management accounting was not even an issue within government. In the second place the audit function at the lower levels of government traditionally has an external character. In particular the annual accounts are an object of audit, directly and exclusively meant for the accountability and discharging process.

Another task inherent factor that causes differences is the place of government in our society. This has to do with the fact that, not entirely unjustly, the citizen looks on government as something which belongs to him. A commission of the Dutch professional body of accountants, NIvRA (1969/ 1970) stated that the citizen wants to know what has happened to "his" money (considered by this commission to be the explanation for the phenomenon), that from the viewpoint of the citizens, fraud in a company seems to be more acceptable than fraud in government. If fraud has happened in a company it is supposed to be the entrepreneur's own fault. The person who has to bear the consequences of the fraud and the person who has failed to prevent it are one and the same, so there is no conflict with the citizen's feelings about justice. If fraud has been discovered in government, the citizen considers himself to be among those who have to bear the consequences, because less money is left for the creation of utility. In this case the person who has failed to prevent the fraud and the person who has to bear the consequences are not the same, so the citizen's feelings about justice are violated. In my opinion it is in these circumstances that we find one of the reasons that criteria for materiality may not be borrowed from the private sector, without further study. The effect of sociopsychological factors is seen within government organizations, but it also surrounds them.

A third task inherent factor comes from the importance of regulations and their usefulness. In this context an utterance of the deputy secretary general of the Dutch Ministry of Welfare, Health and Cultural Affairs is interesting (van der Steenhoven 1988):

> Under the influence of all kinds of qualitative criteria which politicians introduce in laws and regulations, rather a lot of difference of opinion can arise about the exegesis of laws. Simplicity and clearness in legislation should provide the answer, I would like to say in

imitation of Geelhoed and Hirsch Ballin. But politicians are not always able to realise this. Where political compromises have to be made, the accountant's priorities will be put after those of the legislators. Hence also after those of colleagues in the ministry who have elaborated the legislation. But—allow me to reassure you—it is not my opinion that such a working method should be opted for too soon. If something is to be regulated, it should be practicable and auditable. Too often I come across regulations which do not regulate anything but only contain intentions. Then false expectations are stirred up, nothing is sanctioned, because there is nothing to sanction, and decisions are nevertheless taken with a large degree of arbitrariness. In short: administrative chaos is thus generated by legislation.

I think this utterance important because of the unsuspected source from which it comes and because it reflects reality so strikingly. Unclear legislation leads to misuse, improper use, and incorrect use of regulations. The impotence or the feelings of uncertainty with regard to insufficiently tight regulations are exported too easily and too often to the accountant by asking for an auditor's report. Fortunately this point is getting more and more attention. The experiments with auditing schedules described by the legislator who asks for a special auditor's report are useful in several ways. They make it possible to anticipate the possibilities and the impossibilities which an accountant will encounter in a special regulation and thus to narrow the expectation gap. From the quotation it can be seen that the problem can be tackled, but it will surely not be removed.

CONSEQUENCES FOR EDUCATION

From what precedes it can be seen that the environment in which government accountants do their work, as well as the content of their assignments, are of a very special character. Therefore the conclusion is justified, that the time is ripe to allow government auditing to attain academic maturity. Van der Hooft (1985/1986) stated that the average professional baggage which a course in auditing provides is not sufficient without supplement, and that the experience gained in other sectors, for instance in the audit of the financial statements of a company, cannot be applied in the public sector without adaptation.

It is odd that a need for education, which originates from the fact that the objects of audit have to cope with problems of allocation and choice, is now causing a serious problem of allocation. Great difficulties, under a given length of the training period and a standardized student load, are caused by needs for additional education. Something extra in at one end means that the same quantity has to go out at the other. This leads to the question of whether the educational program contains enough "dead wood" (at least in relation to government) which could be cut away in order to create space. In Bac (1976) it has been pointed out that municipal governments engage in activities which cover nearly all types and use nearly all legal forms. This conclusion may be extended to provinces and the national government. Even if one were willing to blame the damage on the horizontal mobility of accountants, still the circumstances under which the government accountant must operate make it unlikely that enough redundant elements in the educational program can be detected. Moreover, it may be clear that both van der Hooft and I set great store by the importance of a widely employable accountant who is a member of one professional body. This should not lead to an aversion against change, but it should be borne in mind that prudence is the mother of wisdom, which suggests a universal basic educational program, as well as an additional specialized applied program.

The content of this specialized program is a function of the preceding considerations. A need for education directed at the acquisition of sufficient knowledge and comprehension of the environmental factors valid to audit government organizations, can be derived from the four rationalisms identified. Existing educational programs are modeled basically on the private sector, where the technico-scientific and economic rationalisms dominate. The other rationalisms need further coverage in the training of a government auditor. This concerns especially the political rationalism which is covered by the theory of public administration, and the juridical rationalism, which is the subject of constitutional and administrative law. I also drew attention to schizophrenia in the economic rationalism of governmental institutions, which is a consequence of the fact that not only micro economic but also macro economic aspects play a role there. This fact and the importance of the budget mechanism create a need for studying economics in the area of public finance. Tilburg University has such a program.

AREAS OF RESEARCH

In this virgin area many topics are worth being researched. A first inventory can most simply be made along the lines of the principal subjects of the educational program of registered accountants, which are:

1. administrative organization and information technology,
2. external reporting, and
3. auditing theory.

I will confine myself to one example from each of these areas.

A major problem in governmental organizations, which has to be solved in the field of administrative organization and internal reporting, is the lack of methods and techniques for the control of processes, the outcome of which is of an abstract character. Most policy goals of governmental bodies are not simply translatable into operational and measurable terms. The draft of a new municipal law charges the municipal council with the obligation to arrange and to maintain an administrative organization, which meets the demands of efficiency and auditability in order to support management adequately. Here the government accountant, with his expertise, is the obvious advisor. This development requires research.

With regard to external reporting, the problem of valuation is an important field of research. A balance sheet in governmental accounts is a rather recent phenomenon. Valuation problems previously played a role in only crown companies and the like, but are now emerging in other contexts. We need to answer the question: what constitutes capital in a government entity? Busines economics uses expected profit as a basis for the value of a company, but what is the value of an entity with a completely different economic philosophy? Income acquisition does not take precedence, but rather a spending of the imposed resources in the most useful direction. A clear philosophy of valuation for the balance sheet of a government entity has not yet been formulated. As yet the balance sheet is no more than the closure of the accounts, which leaves on the balance sheet the footprints of the past. Examples are:

- depreciation for other than microeconomic reasons,
- investments which were charged to reserves,
- subsidies which were used to write off investments, and
- infrastructural provisions covered by the proceeds of land sales.

An important problem is how to cope with such an inheritance. Should the book values, which are missing from the balance sheet, be reconstructed to constitute municipal capital (a term which I hesitate to use) or should provisions be made for future replacements? The yearly burden would then be brought in conformity with the actual operational infrastructure of government activities, and a fairer view of budgetary capacity would be presented. The effect, however, in numerous or almost all cases will be a deficit, and deficits are not acceptable in the domain of lower government. The tempting perspectives of a repressive budgetary supervision regime are then unattainable for the Dutch provinces and the municipalities. Some type of financial guardianship threatens if extra resources are asked from the so-called municipality fund. So, formal legislation is not yet suitable for and certainly does not stimulate an adequate solution of the valuation problems described.

Finally, research into the audit itself, which is the third area in which the accountant has outstanding expertise. An important topic suggests the materiality criteria which are to be applied to government auditing. As I have indicated above, sociopsychological factors make it impossible to borrow the criteria of materiality from the private sector. Further, materiality is affected by the characteristics of an income distributing organization, where the significance of an individual item in the annual accounts is completely different from a firm in which profit results are the critical factors of success and the individual items of the annual accounts viewed in relation to these.

CONCLUSION

I have examined whether a basis exists for reaching harmonization of private and public accountancy and auditing or if there are reasons to let government accounting find its own way to adequate reporting. I have concluded that government auditing is special enough for separate academic specialization.

REFERENCES

Bac, A.D. 1976. *De accountants(controle) in de gemeentelijke huishouding* [The auditor and the audit in the municipal organization]. Rotterdam: Accoord, Erasmus Univer siteit.

Center for International Accounting Development of the University of Texas at Dallas. 1987. *Accounting and Auditing in the People's Republic of China.* Richardson, TX: Council of Europe.

Council of Europe. 1985. *Local Authority Accounting in Europe.* Strasbourg.

Encyclopedie van de Bedrijfseconomie [Encyclopedia of Business-economics]. 1970. deel 5 Controleleer. Bussum: W. de Haan.

Gemeentewet [Dutch Municipal Law].

van der Hooft, W.J. 1985/1986. *Accountant bij de overheid* [The auditor in government]. de Accountant.

Lüder, K. 1989. *Comparative Government Accounting Study, 1989.* Speyer: Hochschule fur Verwaltungswissenschaften.

Ministerie van Financiën [Ministry of Finance]. 1987. *Besluit Taak Departementale Accountantsdiensten* [Resolution on the task of departmental auditing institutions]. Staatsblad 384.

NIvRA, *Controle van overheidsrekeningen* [Audits of Governmental Accounts]. 1969/1970. Report of the Commission on Professional Rules on some problems which are important when auditing governmental accounts. de Accountant.

Provinciewet [Dutch Provincial Law].

Snellen, I.Th.M. 1987. *Boeiend en geboeid* [Captivating and captivated]. Alphen aan den Rijn: Samsom H.D. Tjeenk Willink.

van der Steenhoven, K. 1988. *Nog meer controle zal financiële problemen ministeries niet oplossen* [More auditing would not solve the financial problems of the minis tries]. Staatscourant.

Stevers, Th.A. 1979. *De Rekenkamer* [The Court of Audit]. Leiden/Antwerpen: H.E. Stenfert Kroese.

Vereniging van Nederlandse Gemeenten [Association of the Netherlands' municipalities]. 1988. *De rekening als beleidsinstrument* [Financial reporting as a policy making instrument]. VNG Groene Reeks.

de Vries, J. 1985. *Geschiedenis der Accountancy in Nederland* [History of accountancy in the Netherlands]. Assen/Maastricht: van Gorcum.

THE "USER" ORIENTATION OF PUBLIC SECTOR ACCOUNTING STANDARDS:

AN INTERNATIONAL COMPARISON

K.A. Van Peursem and M.J. Pratt

ABSTRACT

A classification theory of user needs is deduced from the users explicitly identified within international public sector Professional Standards. The theory is applied to identify and compare the current orientation of public sector Professional accounting standards in Australia, New Zealand, the United States, Canada, the United Kingdom, Hong Kong, Singapore, and with reference to the International Federation of Accounting Committees (IFAC). It is concluded that the standards are oriented toward different user groups in the different countries.

INTRODUCTION

A considerable degree of diversity exists in public sector accountability reporting. This diversity is the result of a plethora of piecemeal government,

Advances in International Accounting,
Volume 5, pages 289-312.
Copyright © 1992 by JAI Press Inc.
All rights of reproduction in any form reserved.
ISBN: 1-55938-415-8

tax, and provider-inspired regulations, many of which require financial information on a particular segment or activity within the larger entity. The effect is at best confusing, and at worst inhibiting to the general reader's ability to understand a public sector organization. All in all, the opportunity for comparison and evaluation is highly constrained.

It is in this environment that the major English-speaking professional accountancy bodies have taken the plunge and, over the last several years, begun the slow painful process of developing standards for public sector accountability reporting. Initiated at different times and in varied economic, political, and social cultures, diversity between them might well be expected. Yet a fundamental aim of providing useful information about the entity to the public at large can be presumed to lie at the heart of all these public sector reporting standards.

The *aim* of this paper is to analyze the public sector professional accounting standards in New Zealand, Australia, the United States, Hong Kong, Singapore, the United Kingdom, Canada, and the international public sector standards in order to compare the nature and extent of current reporting practice diversity.

The *objectives* of the study are (1) to develop a fundamental criterion for comparison, (2) to apply the criterion to each country's set of general-purpose professional accounting standards, (3) to identify and conclude upon both the internal consistency (or inconsistency) of each set of standards, and upon the diversity between the country's standards, and (4) to identify the group or groups of stakeholders to which each of the sets of public sector accounting standards appears primarily to be addressed.

The professional standard-setting process for western cultures has seen growth only within the last fifteen years or so, and the number of standards produced for the public sector in each country is markedly less than the standards produced by the same country for the private sector. If an information economics approach is taken and the growth in standards is considered to be a reflection of changing demand, then it can be surmised that the profession's growing involvement in public sector standards is a response to an increasing call for accountability by the public at large. The profession, in its effort to develop standards, appears to be struggling not only with decisions that relate to the treatment of specific public sector transactions but also with the need to establish the very foundation of financial reporting for this unique area. It is believed that the information provided in this comparative study might alert standard setters to potential opportunities or barriers to their own standard setting process, and/or the results may assist in the establishment of new standards.

No academic agreement has been reached as to the public sector's reporting role as distinct from the private sector (Van Peursem 1990). A careful review of the profession's handling of the differences in their own reporting frameworks might provide both clues to that distinction, and further the debate toward a common agreement.

SIGNIFICANT PRIOR RESEARCH

Standard-setting organizations have in all probability reviewed existing public sector reporting standards in their own exploratory efforts. The Australian Accounting Research Foundation indicates as much in their proposals (Greenall, Paul, and Sutcliffe 1988, 1-20). This study is distinct from similar studies, however, in so far as the aim of it is not to produce new standards but to examine and evaluate the focus of those that presently exist.

Lüder (Chan and Jones 1988, 82-105) has completed a comprehensive comparison of government reporting practices (a broader comparison than that provided here) in western European countries. It is intended that the current study will compliment the data base which Lüder has painstakingly collected.

Chan and Jones (1988) edited a collection of essays, subjective in tone, each addressing relevant government reporting issues in the United States, the United Kingdom, Eastern European nations, China, and developing nations. The tone of their effort hints at an urgency on the part of the academic community to understand more fully current practice and related issues in governmental reporting on an international scale. This study is developed with the aim of providing a further contribution to that effort.

METHOD

The following method is applied to achieve the aims and objectives of the study. Background information is obtained by reviewing the AICPA Accountants' Index and selecting articles relevant to the content of public sector standards, and the societies and government agencies which establish them. This is followed by a review of the reporting standards themselves.

A "user" perspective is taken in order to identify the focus adopted (implicitly or explicitly) by the standard setters. This involves identifying the needs of the intended "user" and matching those needs to that required by reporting standards promulgations. It is acknowledged however that actual "users" may be a subset of a broader group of stakeholders to whom accountability reports are owed (Van Peursem 1990) and in fact, some "users" as defined by the standards may be more *accurately* termed "accountees," because they will seldom if ever, actually use the accountability reports. Although arguably more accurate, the term "accountees" is not used because of the prevalence of the alternative term "user" in the standards' phraseology and in the related literature.

The information needs of the user are seen to be an appropriate basis for representing the focus of the standards for a number of reasons. First, the identification of the user is found in every conceptual framework produced by the standard setters, and thus provides a ready basis for comparison. Second, its location within each of the frameworks gives an indication of the importance which is placed on the issue by standard setters. Finally, a previous study revealed that a substantial amount of research has been conducted in the area of user or "accountee" identification (Van Peursem 1990), a fact which highlights the importance with which the academic as well as the professional community treat this topic. The following procedures are taken to complete the study:

1. Potential information "users" are explicitly identified within the body of the standards. These users are listed and classified by virtue of their relationship to the entity.

2. The standards are examined in order to identify the resolutions that are made on basic accounting principles as they apply to public sector entities. For example, some or all of the entity, revenue recognition, going-concern, or historical cost principles as applied in the private sector may be deemed appropriate for the public sector, and are therefore either included or rejected in the public sector promulgations.

3. Each of the standards is analyzed in respect of each of the principles in (2). above to determine if they are appropriate to user needs. In other words, were a (specific) information user sufficiently sophisticated to select the optimal decision for his/her own benefit, would she/he accept or reject the stand currently taken by (each) of the private sector accounting principles? and what application of (each) principle might best satisfy that user's information needs?

4. Based on this analysis, a matrix of user needs matched to appropriate accounting principles designed to meet those needs is prepared. The matrix is then utilized to identify the user who, by implication, is the subject of focus as identified by the way in which each accounting principle is applied in the standard. For example, if a public sector standard advocates the accrual basis of accounting to match costs to revenues, the user who would find this policy most useful may be one who has more interest in the efficiency of operations than the nature of cash expenditures. The objectives which justify the method are explained later in this paper and are illustrated in Figure 1.

Once all the "implicit" users are identified, a review of the results should reveal whether:

●   the nature of the implied information user is consistent (or not) within the standards of each professional body (this will provide an indication as to whether or not the standards are internally consistent);

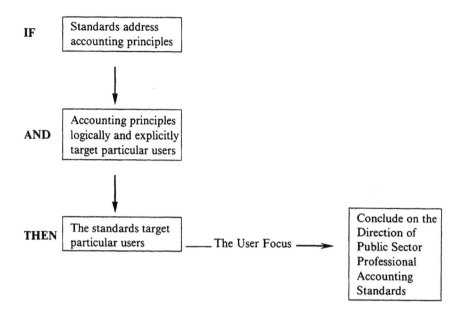

Figure 1. Objectives of the Study

- the implicit and explicit users are (or are not) the same (this will provide an indication as to whether or not the standards promulgated are consistent with the aims established for them); and
- any comparison of focus can be drawn between and among the standards promulgated by different countries.

This process should have the effect of taking the analysis of public sector professional accounting standards beyond that which the standard setters *claim* to accomplish to what they actually accomplish.

THE PUBLIC SECTOR DEFINED

The public sector is defined for the purpose of this paper as being composed of those organizations that receive their authority and/or mission from the government. This group is defined variously in the standards themselves as "government" (CICA 1986, 104), "state and local governmental units" (Bailey 1989, 101), "central government departments of state and local authorities ... majority owned by central government or local authorities ... public education authorities ... any entity established by an Act of Parliament" (NZSA 1987a, 31).

The public sector is viewed here to be inclusive of all the entities within these definitions. The definition assumed for purposes of this paper is inclusive of government business enterprises and service activities although it is recognized that standards for public sector enterprises are frequently the same as for private sector businesses, and a primary focus is placed on the nonbusiness government organization.

A COMPARISON OF THE STANDARDS

Generally accepted accounting principles such as the entity, monetary unit, and revenue recognition principles come under a new form of scrutiny and are even challenged by the accounting standards that are currently under development for the public sector in New Zealand, Australia, the United States, the United Kingdom, Canada, and internationally. (Neither Singapore nor Hong Kong professional associations have issued any public sector standards or guidelines.) They are raised as issues and occasionally resolved, if only in theory, within the framework of an individual standard, and most accounting principles addressed in the standards can be considered to fall among one of the following classifications:

- the scope or nature of the public sector reporting entity;
- the monetary unit principle as applied to the public sector;
- the disclosure principle as applied to the public sector;
- the historical cost principle as applied to the public sector;
- the revenue recognition and matching principles as applied to the public sector;
- the intended users of public sector accountability reports.

The means by which each set of professional accounting promulgations address these principles will be considered below. A summary of each country's professional standards classified by topic is included in Appendix A of this paper.

The Reporting Entity Issue

Judging by the recurrence of the entity identification issue within the earliest of public sector professional standards, the difficulties encountered in resolving the issue may be a major feature that distinguishes the public sector standards from its "parent," the private sector. The entity issue takes many forms but is present in all frameworks, initial standards, or guideline studies. The New Zealand Society of Accountants (NZSA) approaches the issue only indirectly and in a cursory fashion by distinguishing "service" activities from commercial activities and in requiring consolidation (of unspecified entities) for reporting purposes (NZSA 1987a, 52). Forthcoming guidelines may provide further distinction.

J.P. Roy, the Canadian manager on the public sector Accounting and Audit Committee (PSAAC) project advocates a broad definition of the reporting entity by recommending that it be based on three characteristics of the accountee or user: accountability, ownership, and control (Roy 1988, 43). This view was expressed in Canada's third statement of objectives on the scope of the reporting entity (CICA 1986). In this light, it may be seen that an entity need not only be defined by the legal structure or ownership interests (as prevails in the private sector) but by managerial function or by the group or groups within society to whom accountability is owed.

A related topic, consolidation, is also treated by Canada's Statement 4 where the central government "entity" for consolidation purposes is specifically defined to be inclusive of government departments, special funds, and organizations that exist to provide services to the government (CICA 1986, 18). The U.K. position as stated in the Chartered Institute of Public Finance and Accountancy (CIPFA)-prepared Statement of Recommended Practice (SORP) (CIPFA 1987, 3), and the Australian position as stated in the Statement of Accounting Concepts (SAC)-2 (AARF 1985, 32) identify the user need (and the entity) in like manner.

The U.S. Governmental Accounting Standards Board (GASB) and National Council (previously Committee) on Governmental Accounting (NCGA) conceptual statements presently take quite a different track (Bailey 1989). Entity definition is subsumed within the definition and requirements of the long-standing basis of U.S. public sector accounting—"funds". In fact as listed by Bailey (1989) six of twelve major elements of NCGA-1, two statements (6 and 2), one GASB interpretation (1), and two NCGA statements (2 and 7), relate to the reporting of funds (see Appendix B).

This has resulted in an information system that uses the modified cash basis of accounting and aggregation of dissimilar funds, for memorandum only, to report on resource expenditures; in other words, an entity definition which in effect responds to those who wish to know whether compliance with fund expenditure has been met. This has been justified on behalf of taxpayers who according to the concept of "interperiod equity" have a right to know how their current tax dollars were expended in the current period.

The International Federation of Accountants (IFAC) does not address the issue but the narrow scope of their guidelines to business enterprises only must be recalled (IFAC 1989). Australia's ED46A (AARF 1987) and Canada's public sector Accounting Standard (PSAS)-4 (CICA 1986) have approached the issue from quite different perspectives. They do not share, as indicated by the nature of their descriptions, any commitment to the U.S. inter-period equity concept. CICA (1986) identifies the (nonenterprise) entity by the nature of its control of and ownership over resources. Under this view, the entity is composed of resources, and the activities related to them, which are controlled and/or owned by the common source. In the process of identifying each entity for the purposes of inclusion within a particular statutory authority (AARF 1987), Australia recommends that control and influence be among the criteria used to make this distinction. Control and ownership are used therefore under both the Canadian and the Australian models.

Defining and reporting for the government entity appear to be major controversial topics in public sector standards distinguishing them from the equivalent issues in the private sector. They are also distinguished in the frequency of their appearance within the extant professional standards. It remains to be seen whether standards will grow to a point of general agreement on this issue in the future.

Monetary Unit of Measure

Some long-standing accounting principles have achieved wide acceptance through years of use. Practitioners and academicians may find their complacency severely challenged, however, by the public sector environment and its unique accountability needs. The use of the monetary unit to measure the accomplishment of financial and management goals may be one of these issues.

This principle is the subject of renewed attention in some of the earliest and most fundamental standards that have emerged from most of the countries considered by this study. It arises in the form of a renewed interest in the production and report of nonfinancial quantitative and qualitative information as well as financial information.

Conceptual framework standards or guidelines have in fact spoken of the need for nonfinancial information in, for example, Australia (AARF 1987), New Zealand (NZSA 1987a) and 1987b), the United States (NCGA 1980; GASB 1984) and the IFAC (1988, 1989). Subsequent and specific recommendations vary however on the issue as demonstrated by the difference between the U.S. and the New Zealand approach.

New Zealand's PSAS-1 (NZSA 1987b) recommends the production of some statements that may be prepared using no or minimal quantitative material. Witness, for example, their statements of objectives, performance, and resources—all unique to public sector reporting. GASB's (1987b) concept statement, while recognizing the relevance and validity of including qualitative data, makes no attempt to provide a specific guideline for preparation of this information and goes as far as to acknowledge the inadequacy of available non financial performance measures.

CICA (1986) in PSAS-2 (Objectives of Government Financial Statements, p. 17) recognizes that the relationship between revenues and costs in any given period is different in government. It is thereby implied that the information provided by the statement of income would be of little value. Suggestions for its replacement as a surrogate for efficiency are not, however, forthcoming. One can only assume that Income Statement production is encouraged for "special users" (p. 7) and is thereby not under the auspices of this authority. Nonfinancial information is not therefore approached as a topic by CICA.

All in all, the need for qualitative information is generally acknowledged in professional standards for the public sector. However, attempts by the United States and Canada have stopped short of making any recommendations in this regard. New Zealand promulgators may have, by providing alternative formats and by encouraging the presentation of non quantitative information, exposed themselves to greater criticism by doing so in order to provide what may well be more relevant information for the measurement of economy, efficiency, and effectiveness. Not yet resolved in the NZSA standards is either the means for achieving nonfinancial measurement or its report format.

The Disclosure Principle

A wide variety of topics can be classified under the term "disclosure" including, but not exclusively, policy disclosure and consistency, detail of major or unusual transactions, means of aggregation or disaggregation on the face of the statements, and details of litigation in process. A number of public sector accounting standards specifically address the policy and consistency issues. Disclosure requirements for these issues appear in the U.K.'s franked SORP (CIPFA 1987) applying to SSAP-2, "Disclosure of Accounting Policies," in Canada's PSAS-1 and 3 (CICA 1986) and in New Zealand's PSAS-1 (NZSA 1987b, 417). Other disclosure issues are found in the U.S. GASB-3 (GASB 1986) on the disclosure for credit and market risk and NCGA-1 (NCGA 1980) on terminology and classification disclosure issues. Australia's SAC-1 (AARF 1985, 14) has built the disclosure issue into its framework of reporting.

The Historical Cost Principle

A number of standards address asset valuation. The U.S. NCGA-1 (NCGA 1980) does so by recommending the use of historical cost or market value where donated assets are concerned. Canada has no statement on this yet, although a CICA research study (Bloom 1984, 21, 24) indicates a similar stance on the valuation of donated assets. Australia has issued an exposure draft (ED 42C) (AARF 1987) which broadly defines the criteria for the recognition and measurement of assets. New Zealand does not have a separate statement of assets as yet, but does refer to the appropriateness of current cost adjustments in the PSAS-1 (NZSA 1987b, 42) and infrastructural assets (NZSA 1987b, 414). The U.K. profession demonstrates little support of the Current Cost method for this sector (CIPFA 1987, 24). In general, where standards specifically address the valuation of assets, a primary concern (which may be unique in the extent of its application to the public sector), appears to be the valuation of donated and infrastructural assets. Both their valuation and classification are at issue.

Matching and Revenue Recognition

Several standards address the impact of accounting policy on the statement of income. Some are general in nature and encourage the use of either the accrual or some other basis of accounting; others are quite specific to a particular transaction. The accounting treatment of government grant and/ or tax revenues is addressed in the U.K.'s franked SORP on the application of SSAP-4 (CIPFA 1987) Accounting Treatment of Government Grants. Australia's Exposure Draft 46B (AARF 1987) has developed a set of criteria for recognition and measurement of expenses based on probable consumption and reliable measurement.

Depreciation is the issue which seems to attract the most persistent dialogue and yet it receives inconsistent treatment among the different standards. New Zealand's PSAS-1 (NZSA 1987a, 414) in stating that it is inappropriate to depreciate infrastructural assets, accepts that maintenance is a surrogate for depreciation. The U.S. FASB-93 (FASB 1987) takes a different view and recommends that depreciation be charged to income on the depreciable basis of donated as well as other assets. Some are against this position in good part because of application problems that would result and there is evidence that a lobbying effort against this statement is quite active (Anthony 1989).

The promulgators of Canada's PSAS imply that depreciation should be charged to income, as in business, over the life of the asset (CICA 1986, 66) while in contrast, the UK's franked SORP as applied to SSAP-4 (Accounting Treatment of Government Grants) (CIPFA 1987) recommends that local authorities (to which the SORP applies) charge the cost of capital to income over the finance period; depreciation would not be recorded. The argument for this approach is that because the financial position is not dependent on recovering costs through revenues, then depreciation costs (over the asset life) are irrelevant.

Also in SSAP-4 (CIPFA 1987) one finds the recommendation that where grants are received specifically for capital assets, their cost should be included in income over the period of their finance. This relates the revenues to the period of time when funding must be obtained through the imposition of tax levy. Such an approach signifies a break from the revenue recognition principle where recognition relates to the point of earnings. Grants and taxes are not, in effect, "earned" and it is reasonable to assume that the private sector recognition criterion may not be appropriate for this situation.

CICA in PSAS-3 advocates the accrual basis of accounting (CICA 1986, 71) as is implied in Australia's Exposure Draft 46b. The U.S. GASB (GASB 1987a, GASB 1987b) in contrast has used the fund-based approach where government funds are kept on a modified cash basis of accounting. New Zealand's NZSA (1987b) advocates the accrual basis of accounting for public sector entities as well, while the U.K.'s position differs by recommending that asset costs be matched to finance periods (CIPFA 1987). Minimal agreement is found among the different standards therefore with regard to revenue recognition and, particularly, the allocation of asset costs against revenue.

A CLASSIFICATION OF USER NEEDS

Each set of standards provides a list of intended users which are found in the fundamental conceptual standards or guidelines of each set of standards. Their predominance and prominence thus establish the issue as one which is perceived to be important in the eyes of the promulgators.

A number of descriptions are found for the term users. New Zealand's NZSA directs its standard to appease "citizens" including electors, taxpayers, ratepayers, recipients and "resource providers," and all of their representatives including "the media" (NZSA 1987a). In CICA's Statement 3 can be found "the public, legislators, investors, analysts" (1986, 9). Australia's SAC-1 (AARF 1987) identifies users as "providers of resources or their representatives," "service recipients," and "reviewers" such as members of parliament. GASB, in contrast, focuses on present and potential resource providers (Bailey 1989, 215).

What do these "users" share in common? The following section will be dedicated to answering this question and, by so doing, provide a basis for analyzing the direction which each set of standards, taken in whole, may be pursuing. It is argued that any user listed in these standards fits into one of three categories classified by the nature of the user's relationship to the public sector organization: performance evaluators, resource providers, or service recipients.

Reviewers including legislators, investors, and analysts all have in common an indirect relationship to the government entity's management (see Van Peursem 1990 for a description of the indirect relationship) and are forced to analyze from a distance the past efficiencies of managers who are responsible for input resources. As their indirect relationship indicates, their role is to observe, analyze, make investment decisions, or pass on their conclusions to others who do so. They will be classified as those whose primary concern is with management's performance in conducting efficient operations in order to receive a "return" (monetary or otherwise) on their investment. They will be termed *performance evaluators*.

Taxpayers, electors, and their representatives as a group also contribute financial resources to the organization albeit involuntarily through rates, sales, income, and other imposed taxes. Logically, they would be concerned with the use to which those resources are applied (or, the "economy" of purchase) and less concerned with a "return," as any personal return would be only coincidental and not directly matched to their investment. Investors are not included within this classification because, although they provide financial resources, their interests are seen to lie more with the characteristics of efficiency in operation and return on investment which are associated with the performance evaluators. The taxpayers have a primary concern therefore with the economy of input purchases and with compliance with expenditure restrictions. They will be termed *contributors*.

The third category of information user is the *service recipient*. He/she is in an ideal position to place judgment on the quality and effectiveness of the service that the government organization purports to produce by virtue of his/her use of the service. This category would include the health care patient, the ratepayer, the utility user, and the entrepreneur who is affected by trade legislation. The service recipient's primary concern is with the effectiveness of the program (in relationship to his/her special interests); effectiveness may not be measurable in financial terms. They may or may not "pay" for the services but, in all cases, the user fee would be seen to be insufficient to cover costs and the primary concern would be with the quantity and quality of the output.

The fact that these users may be seen to be independent of each other is not to imply that one person may or may not be both a resource provider and a service recipient, a contributor and a resource provider, or a service recipient and a contributor. The assumption is made in this case that they would be motivated to use accountability reports with more than one motivation. Note that, for purposes of this paper, the strength of their motivation is not at issue, but such an issue would be a logical follow-up to this study. A summary of the classification of information users can be found in Figure 2.

Each user as identified by the standards can be classified, it is suggested, into one of these three categories. A number of these users are named within the body of the standards and in fact two of the terms are derived from the Australian SAC-1 (AARF 1987). All user categories specified by the various standards are classified for purposes of this study as either a performance evaluator, a contributor, or a service recipient (see Table 1), based on the principal characteristics of each category (it is acknowledged that any individual could potentially fall under any one or all of these classifications).

A group such as representatives or legislators or media may at any one point in time act as an advocate for any one of the three user groups. Such advocacies would depend on that particular group's ability to purchase or otherwise capture the representative's attention. For example, consumer lobbies (service recipients), taxpayer lobbies (contributors) and municipal bonds rating services

User Characteristic(s)	Termed a
Resource Provider	
Voluntary (Investors and their Representative)	Performance Evaluator
Involuntary (Taxpayers)	Contributor
Service Recipient	Service Recipient

Figure 2. A Classification of User Needs

Table 1. User Classification and Report Functions

	Report Functions to Inform the:		
Country/Organization	Performance Evaluator	Contributor	Service Recipient
New Zealand (SPSAC-4.4)			
Citizens			
Electors	X	X	X
Taxpayers		X	
Ratepayers		X	X
Recipients			X
Resource Providers	X	X	
Representatives	X	X	X
Media	X	X	X
Australia			
Providers of Resources	X	X	
Service Recipients			X
Reviewers (Parliament, Media)	X	X	X
United States (GASB) (Bailey, 1989, 215)			
Present and Potential Resource Providers	X	X	
Canada (CICA, 1986, Statement 3, p. 9)			
Public	X	X	X
Legislators	X	X	X
Investors	X	X	
Analysts	X		
International (Not Specified)			

(performance indicators) may each and at different times act as a primary influence on a local legislation with regard to one local governing authority. Such representatives therefore are classified across all these groups. They could be identified to a particular group if it were known who they were representing, or on whose behalf they were reviewing.

It may be useful to observe that these standards attempt to appeal to a broad group of users and that the classification concepts encompass the primary motivations of all the users specifically identified by the standards.

USER NEEDS AND ACCOUNTING PRINCIPLES

The application of each principle identified as an issue within the standards is now correlated to the motives logically derived for each of the three types of user. This involves postulating the preference which a particular user would be likely to have on each accounting principle.

Six accounting principles were revealed as issues by the nature of the standards. They are: the entity principle, the disclosure principle, the monetary unit principle, the historical cost principle, the matching principle, and the revenue recognition principle. How might each user logically choose to approach each of those principles?

The performance evaluators are likely to be concerned with the efficiency of operations. Their position on the scope of the firm may hold that the entity should consist of those functions under the responsibility of a manager so that responsibility accounting analysis can be adhered to. They might be concerned with input to output measures, whether or not those measures are in monetary units. This is particularly important in a public sector entity, where many of the outputs are in the form of services, not sales. The monetary principle as an exclusive measure of achievement may therefore be challenged by this group. The performance evaluators may find themselves looking for an alternative to historical cost to represent more clearly the value of the assets and the subsequent return on those assets. They may also favor the use of current costs in writing off to income the inflated value of assets purchased in prior years. This viewpoint may also be consistent with support for allocating depreciation to the useful life of the asset in order to match current costs with current revenues or other benefits. Such a principle may also be extended to include donated assets and infrastructural assets in order to more clearly represent the efficiency of their use through the expensing of their value to depreciation.

The contributors may hold quite a different point of view. Their aim may be to be assured that their contributions (usually monetary) are used both for the purposes for which they were donated and otherwise to purchase products at the most economic value. Their view of the appropriate reporting entity may be one defined by the scope of the levy district and the expenditures that issue therefrom. They may be concerned with disclosures that have to do with budget comparison, and sources and uses of finance. Neither depreciation nor current cost methods would be likely to hold any charm for them, as they would be seen to represent artificial, or nonmonetary expenses. The cash basis of accounting would probably be supported, and a list of commitments and future liabilities, may be found useful to ascertain the need for future cash flows.

Service recipients may well define the entity by function. In practice, their interest might be quite narrow and conceivably their involvement with the operations of the entire entity would be of less interest than how they were personally impacted by its constituent parts. But their representatives, such as unions or lobby groups, may take a larger view. Their interests as a group are likely to be in the outcome of service provision and they are likely to be less interested in financial measures. As far as revenue recognition is concerned, they may be most interested in user fees relative to service (accrual) costs. Therefore, they might support the matching concept more than contributors. A matrix that illustrates the applications of likely preferred principles, relative to each users' classification is shown in Table 2.

Table 2. Accounting Principles and Report Functions

	Report Functions to Inform the:		
Accounting Principle	Performance Evaluator	Service Contributor	Recipient
Entity Defined by	Management authority	Levy district or levy purpose	Service function
Disclosure Primary Concern	Input to Output Relationship; ROI; Credit risk	Finance provision; budget comparison	Outcome; user fee relative to output
Monetary Unit	Mixed emphasis	Financial emphasis	Qualitative peformance measures
Historical Cost Position	Current (donated assets) or Historical	Historical	N/A
Depreciation (Matching)	Charged to revenue over useful life	Charged to revenue over finance period	N/A
Revenue Recognition	Accrual; donated assets at market value	Cash basis	N/A

AN ANALYSIS OF THE STANDARDS' FOCUS

The next iteration involves the correlation of a standard's position on each issue with the type of user it implicitly addresses. The process will be advanced by considering the standards of each country in turn. Table 3 matches the six accounting principles called upon or this analysis to the five sets of public sector standards used.

Table 3.   Accounting Principles and User Classification

Accounting Principles	New Zealand	Australia	United States	United Kingdom	Canada
Entity	—	PE/C	C	C	PE/C
Disclosure	PE/C/SR	—	C	PE/C	PE/C
Monetary Unit	PE/C/SR	—	C	—	PE/C
Historical Cost	PE	—	C	C	—
Depreciation (matching)	PE	—	C	CC	PE
Revenue Recognition	PE	PE	C/PE	C	PE

Notes: PE = Performance Evaluator; C = Contributor; SR = Service Recipient; — = Not Clear.

New Zealand

The NZSA is not clear on the entity issue. The only reference to aggregation comes from the commercial versus service sector identification and no analysis on the approach to the entity issue is appropriate at this stage. As to the monetary principle and disclosure issues, PSAS-1 requires, in addition to standard financial statements, the preparation of a Statement of Objectives and Performance budget comparisons (NZSA 1987b) which should appeal to all three user groups.

The advocacy of a current cost method (NZSA 1987a, 54) and the disclosure of ownership of or access to infrastructural assets (NZSA 1987a, 419) favor the needs of efficiency reporting ideal for the performance evaluator. The performance evaluator is also favored by the accrual basis of accounting (following the matching principle) and the charging of depreciable assets to costs of services over their useful life. (Infrastructural assets would not show depreciation as the maintenance is expected to keep pace with its loss in value.)

The NZSA approach shows an orientation, although not exclusively so, to the performance evaluator. The entity issue is not yet resolved through the standard, but should this occur, the New Zealand standards may have quite a clear focus.

Australia

The Australian standards are currently in a formative stage and analysis at this point may be premature, considering the conceptual nature of their current promulgations. It could perhaps be said that a performance evaluator outlook is exhibited in their promulgation of the accrual basis of accounting (ED46b) (AARF 1987) but other issues do not lead to an obvious conclusion. Disclosure issues relate to their general purpose (AARF 1985, 18-22), the monetary unit issue is hinted at in reference to financial performance measures (p. 26). Historical cost is not referred to but a performance evaluator direction is indicated by reference to asset valuation as the "service potential of future economic benefits" (AARF 1985, 54).

The entity issue has been addressed and the standards explicitly target performance evaluators and contributors. The two criteria identifying those to whom accountability is owed (users) are ownership and control of resources. The Australian standards appear to be in a formative stage of development, and effort is being put into a broad foundation prior to the development of specific standards.

The United States

GASB promulgations convey a strong bias toward the contributor user perspective although, as indicated earlier, this view is currently under challenge. The entity is defined by the existence of funds—the backbone of GASB promulgations. Government funds are divided along the line of short-term expenditure periods where the modified cash basis of accounting is primarily used—both being characteristic of the contributor approach. The standard on disclosure of credit and market risk GASB-3 (1986) digresses somewhat as it aligns with the needs of the performance evaluator but other indications favor the contributor.

Depreciation, for example, is not recorded in the major government funds as long-term depreciable assets are kept in a separate fund group. As another example, lip service is paid to the value of nonfinancial information, but no action in the form of specific recommendations for the production of quantitative information follows (Bailey 1989, 206-207).

The United Kingdom

The origin and the existence of the major CIPFA-franked Statement of Recommended Practice (SORP) that has been issued to date suggests the contributor as the targeted user where the local authorities, under the terms of this SORP, are accountable directly to the ratepayer (CIPFA 1987, 1-3). Yet the view that SSAP-12 on depreciation is inappropriate, the emphasis on the disclosure of a rate of return and accrual-based profit, all respond to the needs of contributors or performance evaluaters, rather than service recipients. One finds also that the SORP is not in support of Current Cost Accounting (CIPFA 1987, 24), that depreciation should be on the basis of finance cost (p.18), and that grants are recognized as revenue over the period of finance; all are features which brand this approach as that appealing to the contributor. It should be noted however, that this SORP applies to a narrow range of entities as it is specific to local authorities only. No further indications are evident from what has been produced to date.

Canada

A contributor and performance evaluator perspective is evident from the entity approach taken by Canada's PSAAC. PSAS-2 explicitly attaches input, characteristics of control, use, and ownership of resources to their intended accountees (CICA 1987, PSAS-2, p. 10). It also places substantial emphasis on the standard type of financial statement (CICA 1987, PSAS-3, p. 10) which would appeal to the performance evaluator, yet takes a Contributor viewpoint by promulgating the production of forecast (budget) comparisons (CICA 1982, PSAS-3, p. 97). The production of nonfinancial information is largely ignored (giving less weight to the needs of the service recipient). The accrual basis of accounting is emphasized and a reference indicates that the appropriate method of depreciation is that which charges the cost of assets to revenue over the asset life (CICA 1987, PSAS-3, pp. 66-76). Both of these methods indicate a Performance Evaluation approach. The Canadian approach therefore appears to be a mix between the contributor and performance evaluator points of view, with the service recipient being virtually ignored by the standard policy.

IFAC

It is concluded that the international standards were not sufficiently developed for government service entities to provide a basis for this analysis.

CONCLUSIONS

The user viewpoint adopted by public sector accounting standards, and promulgations in five English-speaking countries, is used to draw conclusions on the current focus and direction of public sector accounting standard setting. The New Zealand standards appear to be both innovative and primarily, but not exclusively, consistent with one set of user goals—the performance evaluation perspective.

It is premature to state at this stage of development whether or not future promulgations in Australia may be singularly focused or will remain consistent with the needs of a performance evaluator as suggested by the conceptual framework. Certainly the nature and extent of their conceptual foundation provides one with the hope that consistency will be attained.

GASB promulgations are more firmly established and are more extensive than any others examined. Indications are that a change is in the offing and this could make irrelevant any conclusions made on the direction they are now and have been taking for many years. It is clear, however, that their focus has been to appease the contributor, despite some confusing input produced by the FASB's conceptual framework.

The Canadian standards may be more readily accepted by the profession in light of their similarity to private sector standards, but a mixed message may be conveyed by the nature of the promulgations. The one clear conclusion that can be drawn on the nature of their standards to-date is the marginalization of the interests of the service recipient.

An active liaison effort may encourage compliance with franked SORPs being issued for public sector organizations in the United Kingdom, but a contributor focus may lead to UK PSACS being regarded as of limited value for management purposes or for the service recipient.

It should not be forgotten that the authority of public sector professional accounting standards is diluted by the impact of statutory disclosure requirements and by the presence of fewer professional accountants within the industry; the sector indeed may be required to rely on the cooperation of and coordination with government and industry.

For these and other reasons it is difficult to assess what direction standard development may take in the future. This comparison illustrates, however, the fact that the standard setters have taken a variety of approaches to meet the needs of accountability and to garner sufficient support to make the standards effective. The U.S. (GASB) standards are probably those most firmly entrenched, but both rumblings within the U.S. academic community and the offering up of fresh alternatives in New Zealand and Australia indicate a dynamic standard-setting environment which may lead to many changes in public sector professional accounting standards.

POTENTIAL AREAS FOR FURTHER RESEARCH

Follow-up empirically based studies would be appropriate to consider the issues listed below.

- Do users' interests tend to fall within the given classifications?
- Do performance evaluators, contributors or service recipients desire the information as postulated by the classification theory?
- What is the accountee response to mixed-user or focused-user reports?
- Can our set of financial statements satisfy the legitimate needs of each of the three user categories?

(APPENDIX FOLLOWS)

APPENDIX A

Summary of Professional Public Sector Reporting Standards by Topic

	Scope and Authority	Conceptual Framework	Disclosure	Entity	Depreciation and Asset Valuation	Matching and Revenue Recognition (Including Pensions, Bases of Accounting)	Users	Liabilities
New Zealand	SPSAC 1.1, 4.5; Explanatory Foreword	SPSAC	Explanatory Foreword, par. 4.17	NA	SPSAC, par. 4.14	SPSAC, par. 4.2	SPSAC, par. 4.2	SPSAC (Commitments)
Australia	SAC-1 par. 3-8; Foreword to SAC and AAS, par. 305.	ED42B	NA	Exposure Draft 42A	ED42B; ED42C	ED46B	ED42A	ED42D
United States	NCGA Interpretation 5; GASB-1	NCGA-1; GASB Concept Statement	GASB-3; NCGA-1	NCGA-3; NCGA-7	GASB-8; NCGA-1	NCGA Interpretation-3; Pensions—GASB 2.4.5; NCGA 4.6 Interpretation-2	GASB Concepts-1	GASB-7; NCGA-1
United Kingdom	SORP; Explanatory Foreword	NA	SORP* no. 2	SORP* no. 1	SORP* nos. 4, 12, 13, 16	SORP* nos. 4, 5, 9	SORP* no. 2	SORP* no. 21
Canada	Introduction to Public Sector Accounting and Auditing Recommendations	IFAC Guideline-1	NA	NA	NA	PSAS-2, par. 39-46 Pensions, PSAS-5	PSAS-2, par. 9-15	NA
International	NA	NA	NA	NA	NA	IFAC-1	IFAC-1	NA

Notes: * The Application of Accounting Standards (SSAPs) to Local Authorities in England and Wales.
NA = No applicable standard.

APPENDIX B

A List of Professional Public Sector Accounting Standards

New Zealand

PSAS-1: Determination and Disclosure of Accounting Policies for Public Sector Service Oriented Activities (1987)

SPSAC: Statement of Public Sector Accounting Concepts (1987); Explanatory Foreward (1987)

Australia

Statement of Accounting Concepts Objectives of Financial Reporting by Public Sector Entities (1985)

Introductory Statement of Applicability of Statements of Accounting Standards to Public Sector Business Undertakings (1985)

AARF: Series No. 2, Exposure Draft 46: Proposed Statements of Accounting Concepts (1987)

Exposure Draft 46A: Definition of the Reporting Entity

Exposure Draft 46B: Definition and Recognition of Expenses

Exposure Draft 42A: Objectives of Financial Accounting

Exposure Draft 42B: Qualitative Characteristics of Financial Information

Exposure Draft 42C: Definition and Recognition of Assets

Exposure Draft 42D: Definition and Recognition of Liabilities

United States

GASB Statement-1: Authoritative Status of NCGA Pronouncements and AICPA Industry Audit Guide (1984)

GASB Statement-2: Financial Reporting of Deferred Compensation Plans Adopted under the Provisions of Internal Revenue Code Section 457 (1986)

GASB Statement-3: Deposits with Financial Institutions, Investments (including Repurchase Agreements), and Reverse Repurchase Agreements (1986)

GASB Statement-4: Applicability of FASB Statement-87, "Employers' Accounting for Pensions," to State and Local Governmental Employers (1986)

GASB Statement-5: Disclosure of Pension Information by Public Employee Retirement Systems and State and Local Governmental Employers (1986)

GASB Statement-6: Accounting and Financial Reporting for Special Assessments (1987)

GASB Statement-7:   Advance Refunding Resulting in Defeasance of Debt
                    (1987)
GASB Statement-8:   Applicability of FASB Statement No. 93, Recognition
                    of Depreciation by Not-for-Profit Organizations, to Certain
                    State and Local Governmental Entities (1988)
GASB Concepts
  Statement-1:   Objectives of Financial Reporting (1987)
  NCGA Statement-1:   Governmental Accounting and Financial Reporting
                      Principles (1980)
  NCGA Statement-2:   Grant, Entitlement, and Shared Revenue Accounting
                      and Reporting by State and Local Governments
  NCGA Statement-3:   Defining the Governmental Reporting Entity
  NCGA Statement-4:   Accounting and Financial Reporting Principles for
                      Claims and Judgments and Compensated Absences
  NCGA Statement-5:   Accounting and Financial Reporting Principles for
                      Lease Agreements of State and Local Governments (1983)
  NCGA Statement-6:   Pension Accounting and Local Governments
                      Employers (effective date deferred)
  NCGA Statement-7:   Financial Reporting for Component Units within the
                      Governmental Reporting Entity (1984)

  NCGA Concepts Statement-1:   Objectives of Accounting and Financial
                      Financial Reporting for Governmental Units

  FASB Statement of Financial Accounting Concepts No.4:   Objectives of
                      Financial Reporting by Non-business Organizations (1980)
                      (Interpretations, Exposure Drafts, SOPs, Discussion Memo-
                      randa not included).

United Kingdom

The Chartered Association of Certified Accountants: Explanatory Foreward
                    paragraphs 17-18 (Public Sector)
(Franked) Statements of Recommended Practices (SORPs) applicable to the
                    public sector: Accounting practices for Scottish Local
                    Authorities—Statements of Recommended Practice, Nos. 1-4
                    (1987-1989), Nos. 5-6 (to be issued)
The application of Accounting Standards (SSAPs) to Local Authorities in
                    England and Wales (1987)
Local Authority Accounting (1987)

Canada

CICA Introduction to Public Sector Accounting and Auditing Recommen-
                    dations (1986)

PSAS-1:   Disclosure of Accounting policies (1983)
PSAS-2:   Objectives of Governmental Financial Statements (1987)
PSAS-3:   General Standards of Financial Statement Presentation for Governments (1986)
PSAS-4:   Defining the Government Reporting Entity (1988)
PSAS-5:   Accounting for Employee Pension Obligations in Government Financial Statements (1988)

International

Introduction to the Public Sector Committee of the Internatinal Federation of Accountants (1988)
Guideline 1:   Financial Reporting by Governmental Business Enterprises (1989)

ACKNOWLEDGMENT

We wish to thank Assoc. Professor H.H.B. Perera for his helpful comments on an earlier version of this paper.

REFERENCES

Anthony, R. 1989. Memorandum for the members of the government and nonprofit section, American Accounting Association. Harvard University Graduate School of Business Administration, Boston.

Australia Accounting Research Foundation (AARF). 1985. *Statement of Accounting Concepts: Objectives of Financial Reporting by Public Sectors Entities*. Melbourne: Author.

————. 1987. *Proposed Statements of Accounting Concepts, Series No. 2-PSAC*. Melbourne: Author.

Bailey, L. 1989. *Miller Comprehensive 1989 Governmental GAAP Guide*. San Diego: Harcourt Brace Jovanovich.

Bloom, R. 1984. American and Canadian standard setting: A comparative analysis. *International Journal of Accounting Education and Research*, 19.

Canadian Institute of Chartered Accountants (CICA). 1986. Public Sector Accounting and Auditing Committee. *Public Sector Accounting Statements One Through Four and Introduction*. Toronto: Author.

Chan, J.L. and R.H. Jones. 1988. *Government Auditing and Reporting*. London: Routledge.

Chartered Institute of Public Finance and Accountancy (CIPFA). 1987. Accounting Standards Committee (ASC) and the Accounting Standards for Local Authorities Group. *The Application of Accounting Standards [SSAPs] to Local Authorities in England and Wales*. London: Author.

Financial Accounting Standards Board (FASB). 1987. *FASB Statement No 9: Depreciation in Nonprofit Accounting*. Stamford, CT: Author.

Governmental Accounting Standards Board (GASB). 1984. *GASB Statement 1: Authoritative Status of NCGA Pronouncements and AICPA Industry Audit Guide*. Stamford, CT: Author.

_____. 1986. *GASB Statement 3: Deposits with Financial Institutions, Investments (including Repurchase Agreements), and Reverse Repurchase Agreements.* Stamford, CT: Author.

_____. 1987a. *GASB Statement 6: Accounting and Financial Reporting for Special Assessments.* Stamford, CT: Author.

_____. 1987b. *GASB Concepts Statement 1: Objectives of Financial Reporting.* Stamford, CT: Author.

Greenall, D.T., J. Paul, and P. Sutcliffe. 1988. *Financial Reporting by Local Governments.* Melbourne: Australian Accounting Research Foundation.

International Federation of Accountants (IFAC). 1988. *Introduction to the Public Sector Committee of the International Federation of Accountants.* New York: Author.

_____. 1989. *Financial Reporting by Government Business Enterprises.* New York: Author

National Council of Governmental Accountants (NCGA). 1980. *NCGA Statement 1: Governmental Accounting and Financial Reporting Principles.* Stamford, CT: Author.

New Zealand Society of Accountants (NZSA). 1987a. *Statement of Public Sector Accounting Concepts.* Wellington: Author.

_____. 1987b. *Determination and Disclosure of Accounting Policies of Public Sector Service Oriented Activities.* Wellington: Author.

Roy, J.P. 1988. Government reporting: The big picture. *CA Magazine* 121. 49.

Van Peursem, K.A. 1990. *A definition for public sector accountability.* Discussion Paper No.105. Department of Accountancy, Massey University, Palmerston North.

Printed and bound by CPI Group (UK) Ltd, Croydon, CR0 4YY

08/05/2025

01864950-0005